DESIGN COMPUTING

Design Computing will help you understand the rapidly evolving relationship between computing, designers, and the many different environments they create or work in. The book introduces the topic of design computing, and covers the basics of hardware and software, so you don't need to be an expert. Topics include the fundamentals of digital representation, programming and interfaces for design; the shifting landscape of opportunity and expectation in practice and pedagogy; placing sensors in buildings to measure performance; and the challenge of applying information effectively in design. The book also includes additional reading for those who wish to dig deeper into the subject. *Design Computing* will provide you with a greater awareness of the issues that permeate the field, opportunities for you to investigate, and perhaps motivation to push the boundaries.

Brian R. Johnson, Associate Professor of Architecture and Director of the Design Machine Group at the University of Washington, is also past president and recipient of two Awards of Excellence from the Association for Computer Aided Design in Architecture (ACADIA). His 35-year teaching career has witnessed the emergence of design computing.

DESIGN COMPUTING

An Overview of an Emergent Field

Brian R. Johnson

NEW YORK AND LONDON

First published 2017
by Routledge
711 Third Avenue, New York, NY 10017

and by Routledge
2 Park Square, Milton Park, Abingdon, Oxon OX14 4RN

Routledge is an imprint of the Taylor & Francis Group, an informa business

© 2017 Taylor & Francis

The right of Brian R. Johnson to be identified as author of this work has been asserted by him in accordance with sections 77 and 78 of the Copyright, Designs and Patents Act 1988.

All rights reserved. No part of this book may be reprinted or reproduced or utilized in any form or by any electronic, mechanical, or other means, now known or hereafter invented, including photocopying and recording, or in any information storage or retrieval system, without permission in writing from the publishers.

Trademark notice: Product or corporate names may be trademarks or registered trademarks, and are used only for identification and explanation without intent to infringe.

Library of Congress Cataloguing in Publication Data
Names: Johnson, Brian Robert, author.
Title: Design computing : an overview of an emergent field / Brian R. Johnson.
Description: New York, NY : Routledge, 2017. | Includes bibliographical references and index.
Identifiers: LCCN 2016020489| ISBN 9781138930964 (hb : alk. paper) | ISBN 9781138930971 (pb : alk. paper) | ISBN 9781315680057 (ebook)
Subjects: LCSH: Building--Data processing. | Architectural design--Data processing. | Buildings--Computer-aided design. | Architecture--Computer-aided design.
Classification: LCC TH437 .J638 2017 | DDC 690.0285--dc23
LC record available at https://lccn.loc.gov/2016020489

ISBN: 978-1-138-93096-4 (hbk)
ISBN: 978-1-138-93097-1 (pbk)
ISBN: 978-1-315-68005-7 (ebk)

Acquisition Editor: Wendy Fuller
Editorial Assistant: Trudy Varcianna
Production Editor: Hannah Champney

Typeset in Bembo and ITC Stone Sans
by Saxon Graphics Ltd, Derby

Printed and bound in Great Britain by
TJ International Ltd, Padstow, Cornwall

CONTENTS

Foreword	*vii*
Preface	*xi*
Acknowledgements	*xvii*
1 Introduction	1

PART I
Starting Points — 17

2	Models	19
3	The Built Environment	25
4	Thinking Inside the Box	34
5	Doing What Designers Do	51

PART II
The Grand Challenges — 75

6	Design Problems: What Are They?	77
7	Cognition: How Do Designers Think?	84
8	Representation: Capturing Design	104

9	Interface: Where the Action Is	127
10	Practice: Data, Documents, and Power	139
11	Expertise: Challenges and Opportunities	154
12	Solutions: Generation and Refinement	162
13	Buildings: Computation Sources and Sinks	175
14	Pedagogy: Teaching the New Designer	184

Index   *193*

FOREWORD

I met Brian Johnson online. The year was 2005, and I had applied for a tenure-track teaching position at the University of Washington in Seattle. In the back-and-forth communications that are an inevitable part of academic job applications, I came to know Brian, and after the search reached its conclusion he encouraged me to join ACADIA (the Association for Computer-Aided Design in Architecture). My hunt for employment eventually took me to North Dakota State University, where from a vantage point some ten years distant, I can look back on my time in ACADIA and on conversations and collaborations with Brian over the years as critical to my understanding of the issues and questions of design computing.

But what exactly *is* design computing? In 2014, I invited Brian, Erin Carraher, Ning Gu, JuHyun Lee, and Wassim Jabi to join me in co-editing a special issue of *Architectural Science Review* on the theme of design computing (issue 58.3). As a group, we worked to define a shared understanding of the rapidly expanding content and the nebulous boundaries of the field. Its disciplinary content is characterized by a range of interests and issues which frequently overlap and cross-pollinate: simulation, visualization, and representation; human–computer interaction, user experience, and cognition; artificial intelligence; fabrication and prototyping; parametric design and scripting; data processing and intelligent modeling; smart environments, controls, and infrastructures; algorithmic thinking; mobile, ubiquitous, and pervasive computing; and cloud computing—just to name some of the most central. Clearly, design computing, which Brian argues exists at the intersection of the built environment, design, and computing, is a content-rich field about which much has been written but which (as of yet) shows no risk of exhausting itself as a robust field of inquiry.

Its boundaries are indeed nebulous and characteristically difficult to define. At times, design computing shares formal methodologies with computer science and mathematics, seeming to succumb almost wholly to algorithmic approaches and the application of logical reasoning. However, considered at other times or looked at in different ways, design computing forcefully calls the value of logic into question: Intuition, aesthetics, and human experience all have their roles to play. To precisely delineate the boundaries of design computing is a Sisyphean task. In one sense, I see Brian's book as an attempt to define the field's core values and approaches through comparison with its kindred disciplines. Often, he makes these comparisons explicit, as when he discusses traditional and digital media; but occasionally the comparisons are latent within his text and emerge only after repeated readings.

Like mathematics, design computing is concerned with patterns, whether those are patterns of thought, patterns of practice, or visual or spatial patterns. In particular, design computing is concerned with the *computability* of patterns—the *how* and *why* of algorithmic approaches. But unlike the rigorous discipline of pure mathematics, at its core design computing is about application—and it is a unique kind of application which demands constant interpretation and self-critique. Brian addresses this relationship in his first chapter:

> In calculus [he writes] we learn that functions have minima and maxima that can be found by differentiating the equation and solving for the derivative's zero values. Where such situations exist, that is, where a mathematical expression describes the "value" of a design, there is a direct mathematical, or analytic means of arriving at the best solution. We can, for example, select the best depth for a beam carrying a given load. Automating the improvement of a design is hard. There may be no clear mathematical relationship between the changes you can make to the representation and the performance metrics you wish to improve.

Thus, while design computing is abidingly concerned with the patterns and structures of computing, the questions that drive the field emerge from the needs of design. Quantifiability and computability are indeed essential tools, but there remains a dimension of design computing clearly inaccessible to automation and algorithms. Now, again like mathematics, and also like industrial or product design, the discipline of design computing seeks to identify and solve problems of a wide range and scope. But as Brian acknowledges, every problem isn't a design problem, and design doesn't require a computer. Rittel and Webber's *wicked problems* (Rittel and Webber 1973) hold special significance in design computing, for like acts of design, these problems are open-ended and resistant to single solutions.

Regardless of their specific discipline, designers rely on media: the material of their work. For example, architects are characterized—at least historically so—by their reliance on drawings and physical models; graphic designers similarly by paper and film; and product designers by physical prototypes. Designers' media are structured—think about how they conventionally adhere to established rules of representation and the limitations of scale, or how the structure of software influences the direction and scope of inquiry. Design computing recognizes that the structure of media, whether analogue or digital, is often computable and thus inviting of algorithmic analysis. And yet again, there remains a dimension to the field which is not obviously susceptible to a pure-mathematics approach. Brian notes how the "fluidity of interaction and range of expression" made possible by pen and paper is substantially challenging to register through digital tools. With this comment and in his subsequent discussion, he suggests that design computing—perhaps more so than other design disciplines—acknowledges that media and representations are not neutral with respect to their application or their outcomes.

Design computing's concern with the built environment extends from inquiry into the processes of architectural design, through the exchange of information between constituents, to the logistics of construction, and finally to the mechanics of responsive environments and the management of real-time information about the operation of built structures. Brian touches on all of these aspects in his book. As a field, building science addresses the performance and constructability of buildings, dealing with measurement, logic, and integrated systems. Design computing shares these concerns, but they are viewed through the lens of computation, which means that building performance and constructability are not considered simply measurable, but that they are susceptible to algorithmic processes. Buildings, as Brian writes, are potentially both computer *input* and *output* devices—operating both to collect data and to respond in real time, whether physically or numerically.

As counterintuitive as it may seem to a reader new to the field, design computing has a rich historical dimension. Except for a handful of writings from the pre-transistor era (e.g. Thompson 1917), few of the seminal writings in the field (e.g., Sutherland 1963; Alexander 1964; Negroponte 1970; Newman and Sproull 1973; Stiny 1975; Alexander et al., 1977; Hillier and Hanson 1984; Mitchell 1977, 1990) date prior to the 1970s. Yet, distinct facets within the wide field of design computing trace their roots to individuals and movements as varied as Babbage, von Neumann, Turing, Herbert Simon, Deleuze, Gestalt psychology, economics, and genetics. Along with an ever-expanding discovery and adaptation of material from historical sources deemed newly relevant, the horizon of design computing is always expanding outward to encompass new technologies. Along with this rapid expansion of design computing have come serious questions about the designer's disciplinary identity. When designers come to rely ever more strongly on computational strategies, and as software takes on an increasing role in the production of designerly objects, legitimate concerns arise about what Brian cites as the "protected core" (Archea 1987): To what agency are we to ascribe a designed work when that work registers the hands not only of illustrators and architects, but also of the team that designed the software?

When attempting the explication of such a diverse and far-reaching field, a grasp of its content and boundaries is necessary. Brian Johnson writes with incisive clarity and expert knowledge of his multi-faceted subject. This alone should make you want to read his book: Navigating the deep ocean of design computing is not a task you should undertake without an experienced guide like Brian. But more importantly, he writes with real joy and enthusiasm and an obvious desire to share that enthusiasm with his readers. For this reason, his book is not only informative but is a true pleasure to read.

Mike Christenson, AIA
Associate Professor of Architecture
North Dakota State University
March, 2016

References

Alexander, C. 1964. *Notes on the synthesis of form*. Cambridge, MA: Harvard University Press.
Alexander, C., S. Ishikawa, and M. Silverstein. 1977. *A pattern language: Towns, buildings, construction*. New York: Oxford University Press.
Archea, J. 1987. Puzzle-making: What architects do when no one is looking, in *Computability of design*. Edited by Y. Kalay, 37–52. New York, NY: Wiley.
Hillier, B. and J. Hanson. 1984. *The social logic of space*. Cambridge: Cambridge University Press.
Mitchell, W. J. 1977. *Computer aided architectural design*. New York, NY: Van Nostrand Reinhold Company.
Mitchell, W. J. 1990. *The logic of architecture*. Cambridge, MA: MIT Press.
Negroponte, N. 1970. *The architecture machine: Toward a more human environment*. Cambridge, MA: MIT Press.
Newman, W. and R. Sproull. 1973. *Principles of interactive computer graphics*. New York, NY: McGraw-Hill.
Rittel, H. J. and M. M. Webber. 1973. Dilemmas in a general theory of planning. *Policy Sciences* 4: 155–169.
Stiny, G. 1975. *Pictorial and formal aspects of shape and shape grammars*. Basel: Birkhauser.
Sutherland, I. E. 1963. *Sketchpad: A man–machine graphical communication system*. Lexington, MA: MIT Press.
Thompson, D. 1917. *On growth and form*. Cambridge: Cambridge University Press.

PREFACE

> May you live in interesting times.
>
> *Ancient curse*

Designers of the built environment, especially architects, are currently seeing the collision of design and computing with traditional built environment challenges and opportunities. New digital tools and opportunities are changing how we design, what we build, how we build it, who uses it, how we do business, and how we learn. Old questions about design process, creativity and problem solving are resurfacing in new forms, and new questions about agency, optimization, and expertise are emerging as well. Whether a curse or a blessing, these are interesting times!

Design is a complex task involving experience, creativity, awareness of material properties and costs, evaluation with regard to both quantifiable and aesthetic aspects, judgment and communication skills, as well as sensitivity and adherence to many legal and cultural standards. Computers are fast and efficient tools for carrying out well-defined (some would say rigid) sequences of computations, preserving and editing large amounts of data, and—when connected in networks—communicating clearly over great distances, often surprising us in the process. Efforts to support, assist, and/or automate design, construction, and inhabitation of the built environment using computers have been going on for several decades, with the result that most students today learn to design on a computer, most large construction depends on computing, and many buildings and environments are digitally augmented in one or more ways. We have available to us many complex software and hardware systems representing thousands of man-years of development effort and demanding hundreds of hours of sometimes frustrating time investment in study and practice in order to master. Our digital tools are opening up new design opportunities, but they are also changing the way we work and the things we work on. Digitally augmented lighting, heating, security, and entertainment systems can be elegantly woven into the built fabric of our lives, or crudely applied as an appliqué rather than an integral part of the design. The human activity of design and the advanced technology of computing exist in a tangle of cause and effect. We create and inhabit designed spaces, deploying and consuming digital technology, controlling and being influenced by software design. It is an exciting time, motivated by deep fundamental questions, but often dominated by daily details. It is easy for the froth of daily technology engagement to obscure the larger picture.

"Design computing" is one of several terms that have evolved to describe the area of study growing at the intersection of computer science, design, and the built environment, often engaging both the detail-oriented specificity of the software engineer and the free-wheeling "outside the box" thinking of the artist, while grappling with the real-world constraints of design and fabrication. If you use software in your design process, or build software for those who do, you are part of this intersection. If your lights are linked to occupancy sensors, your blinds adjusted by lighting sensors, or your thermal comfort entrusted to a smart thermostat, you're also part of this. Understanding the forces operating at the intersection will help you work smarter as a designer, make better choices as a software user, building occupant, or building owner, and deliver a superior product as a building or software developer.

Computer software helps you get things done. Sometimes it is fun to use; at other times it is very frustrating. A lot of time is invested in acquiring or updating computer skills, both by students seeking entry-level skills and by practitioners pursuing advanced opportunities. Yet, there is no certainty that the software you learn this year will be the software you use next year, and even less chance of a match in a decade, given inevitable updates and periodic technology revolutions. It may seem that learning new software is a perpetual and perhaps distracting cost of doing and learning to design. There are many books, tutorial websites, and online videos that aim to help you learn to use a particular program, which is great when that need matches your goals, but they have notoriously short shelf lives, and it may be hard to apply the lessons from one of these to another program. Can anything be done to make your life easier? Are there themes that run through the different programs? What do experts in design computing carry over from one system to the next, or one decade to the next?

Who Should Read this Book?

This is a book for people interested in the big questions that motivate the field of design computing—students and practitioners of design (especially architecture) who want to know more about the technology they use, or computer programmers who want to know more about the activity they support. It is not a book about a particular software package, but about the concepts, problems, constraints, and trends behind computer-aided design software in general. If your experience with computing has been largely about completing tasks for work or school—writing papers, making presentations, drafting or modeling, even learning to program—

you will come away with a deeper understanding of these activities and ideas by reflecting on the grand challenges and tentative answers that researchers and software developers have produced so far.

There are opportunities and challenges raised by these questions. Where both creativity and technology are involved, people often connect well with one domain but struggle with the other. Both are important. This book provides a high-level overview and a starting point from which to explore more deeply. By employing plain English, readers from either side of the line should find value and increased comfort with the other side. The questions are challenging. Some might not even have solutions; the goal is to understand their general form and import, and to understand how they shape current options or actions. Today's systems inevitably address yesterday's understanding and problems. While useful, even "interesting," they fall far short of the ideal. We need to create new, interdisciplinary approaches to solving tomorrow's problems. We need the emerging field of design computing.

Expertise: Operational Skill + Wetware

There are two aspects to becoming a digital power user: *operational skills* and *wetware*. Operational skill arises from working with hardware and software: learning menus, commands, buttons, options, and effects. It is specific to particular programs, devices and even versions of the software. It requires studying and repeating operations to the point where you don't expend conscious thought on their execution. *Wetware*, in contrast to hardware (equipment) and software (programs), describes what's in your head—the conceptual frameworks that you bring to the work (Rucker 1988). It encompasses expectations and learning from previous experience with similar projects and programs, as well as your understanding of the central character of a program, the strategy of its best use, and its limitations. Wetware is the central component of *computer literacy*—the "capability for effective action" in a new situation (Denning *et al.* 1989). While there isn't a substitute for "practice, practice, practice" in developing operational mastery, hand–eye coordination, and muscle-memory (i.e., speed), your understanding and skill in navigating among strategic choices can be enhanced by study and reflection, studying the information-processing technologies that govern systems and reflecting on the information requirements of design tasks, considering those activities that are amenable to some sort of computer assist, augmentation, or replacement.

The concept of wetware reminds us that computing is an ecosystem requiring human understanding, in addition to software and hardware. We have been told, mostly by advertisers, that using a computer is a simple and intuitive activity that anyone can master—just "point and click!" The truth is more complex. Hardware and its features have a lot to do with how productive we can be, but so does the way you interact with the machine, what software you have, and your experience. The software fits between your own needs and the machine's power, but it is your knowledge that guides your actions. If you have a powerful computer with functionally powerful software, but don't have suitable wetware, you may well find yourself lost in a maze of menus, icons, and poor choices.

It may not always seem that way, but software developers do try to create tools that accomplish tasks efficiently, match your skills and expectations, work with available hardware, and use appropriate resources of time and space. This means the software embodies assumptions about the task you will undertake and the steps necessary to complete it. It also means that the software embodies and is shaped by mathematics and practical knowledge; it may do amazing things, but it can't do magic no matter how much you ask. Developing and deepening your understanding of the foundational limitations, assumptions, cognitive frames, and opportunities of design computing is the main focus of this book.

Having detailed knowledge of these topics may not be necessary for all, but will be critical to some, and of interest to many. Clive Thompson, in his book *Smarter than You Think*, suggests that those who understand software, and how best to use it in conjunction with their native intelligence and personal skills, are often able to outperform expert systems (Thompson 2013). In my teaching I call this "working smart"—learning enough about the digital environment and your design process to make effective use of the computer to reach your goals.

How to Use this Book

The book is divided into three parts. Chapter 1, *Introduction*, develops a very high-level sketch of design computing emerging from its three primary constituents: the increasingly complex and digital built environment itself; the humanistic and cognitive traditions of design; and the exacting requirements of computing with regard to logic and mathematics. The interaction of these three influences is traced quickly through the historical evolution of efforts to aid design by applying computation.

The introductory chapter is followed by two, more extensive, parts.

Part I: Starting Points offers four chapters that visit a number of foundational subjects that underlie the rest of the book, including models, computational thinking, and a data-centered overview of the architecture, engineering, and construction (AEC) industry as a whole, before reviewing the current state-of-the-art.

Part II: Grand Challenges comprises nine chapters examining in more detail the various practical and theoretical issues that confront those who seek to apply digital tools to the design domain. The core puzzles represent key motivating questions in the search for useful computational tools. Some focus on what designers do—What is a design problem? How do designers think? What is the role of drawing and model-making? How can we analyze and evaluate designs? How can we improve on an existing design? And how do we best teach designers to use information systems? Some puzzles are concerned with how technology interacts with the practice of design, changing roles and power relationships in the building professions, or how it enhances buildings. Finally, since computing, like paint and glass, is used increasingly in the finished product, designers find themselves confronting the need to deploy and control it in their designs and cope with it in their working environment. Through consideration of these disparate challenges in design computing, both designers and programmers gain a greater understanding of our technology, our products, and ourselves, enabling us to work smarter.

At the end of each chapter you will find suggested readings and references that can provide starting points for further investigation. The literatures on architecture and computer science are vast on their own, and occasionally challenging to the "crossover" reader due to jargon or detail, but the references chosen for this book are generally accessible to any reader and can help you get started on a deeper exploration.

We do live in interesting times, with all the opportunity and perils implied by that "ancient curse." I hope this book helps you to navigate them, and perhaps find some inspiration to dig deeper into design computing.

References

Denning, P. J., D. E. Comer, D. Gries, M. C. Mulder, A. B. Tucker, A. J. Turner, and P. R. Young. 1989. Computing as a discipline. *Communications of the ACM* 32 (1): 9–23.

Rucker, Rudy. 1988. *Wetware*. New York, NY: Avon Books.

Thompson, Clive. 2013. *Smarter than you think: How technology is changing our minds for the better*. London: Penguin Press.

ACKNOWLEDGMENTS

This book project would not have been undertaken without the curiosity of my students. It would not have been informed without the stimulation of my many friends from the early days of the Association for Computer Aided Design in Architecture (ACADIA) and the rapidly expanding field of design computing in general. Finally, it would not have been written without the support and patience of my wife and daughter.

1
INTRODUCTION

> Where there is great power there is great responsibility.
> *Winston Churchill (1906)*

Design computing is just one of several names for the discipline emerging at the intersection of design, computer science, and the built environment. Others, each reflecting a slightly different viewpoint, include *architectural computing* and *computational design*, *algorithmic design* and *responsive architecture*. In this discipline issues of theoretical computability mix with questions of design cognition, and personal design intuition runs head-on into state-space search. The overlaps have spawned new areas of inquiry and reinvigorated existing ones, such as smart environments, pattern languages, sensors and machine learning, visual programming, and gesture recognition.

The study of design computing is concerned with the way these disciplinary threads interweave and strengthen one another; it is not about learning to use a particular piece of software. Of course, along the way you may acquire skills with several pieces of software, because the best way to understand computing machinery is to work with it, but the goal is to master the concepts that underpin the software, enabling you to attack new problems and new programs with confidence and get up to speed with them quickly.

To begin this study, we'll look at each of the three major topics on its own, from the inside, and then consider their interaction; but let us begin by setting the broad cultural context—the outside view—for these subjects.

Design Computing: An Uneasy Juxtaposition

Design and computing are both perceived as moderately opaque topics by many people, for somewhat opposite reasons, and those who readily understand one seem especially likely to find the other challenging. Design is often described using poetic metaphors and similes (e.g., Goethe's "Architecture is frozen music"), while computing is covered with various sorts of obscure detail-oriented jargon and techno-speak (e.g., "The Traveling Salesman Problem is NP-complete"). Both vocabularies, while meaningful to the insider, may leave others feeling cut off. Nonetheless, effective use of computing tools in pursuit of design or development of new digital tools for conducting design requires both kinds of knowledge. In this chapter we will explore the relevant parts of both vocabularies in language that most readers will be able to understand and utilize, setting both against the background of the built environment itself.

In an era heavily influenced by concepts drawn from science, design remains mysterious, often portrayed as producing new and engaging objects and environments from the whole cloth of "inspiration." Within design professions, especially architecture, those who actually do design usually occupy the top of the power structure, in positions reached after many years of experience. Outside of the design professions there is increasing respect for the economic and social potential of design through innovation, creativity, and thinking "outside the box." At the same time, more and more actual design work takes place inside the black boxes of our computing tools. Unfortunately, the fit between the mysterious design task and the highly engineered digital tools is poorly defined, with the result that users often revel in newfound expressivity and simultaneously complain about the tool's rigid constraints.

Design remains mysterious in part due to the way designers are trained—while consistently described in terms of process and iteration, students commonly learn to design by being dropped into the deep end of the pool, told to do design, and then invited to reflect on the activity with a mentor, gradually developing a personal design process (Schön 1984). While that might seem to encourage simple willfulness, you instead find design described as reconciling the seemingly irreconcilable, and resistant to precise definition.

In contrast, computer technology, both hardware and software, is built on an explicit foundation of physics and electronics rooted in physical laws and absolute predictability. They do what we design them to do. Perhaps this means they cannot be creative, as Lady Ada Lovelace and others have avowed (see Chapter 12), but whatever the ultimate answer is to that question, systems have become so complex in their operation as to produce an increasing impression of intelligent creative results. In the end, the challenge seems to be how best to use these highly predictable machines to assist an unpredictable process.

Similarities

Though traditional views of design and computer programming display great differences, they both involve detailed development of complex systems. In addition, computer programming is often learned in much the same way as design—through doing and talking—and computer science has even adopted the

Introduction **3**

term *pattern language*—a term originally coined by architect Christopher Alexander—to describe efforts in software design to name and organize successful strategies.

Design computing is concerned with the interaction of design, buildings, and computing. As illustrated in Figure 1.1, this can be broken down into three bi-directional pairings: computing and design, design and building, building and computing. Each contributes to design computing.

Computers are used heavily as a production technology in architectural practice, offering opportunities to the designer in the form of CAD software and digital media, but there is more to it than improved productivity. Digital tools enable entire new categories of shape to be conceived and constructed, but they also invite architects to participate in design of virtual environments for games and online activities. At the same time, the tools influence relationships within offices. Designing with digital tools is changing how we think about routine design and leading to whole new theories of "the digital in architecture" (Oxman and Oxman 2014).

Computing is already an important component of construction planning and execution, used in scheduling and cost estimating, checking for spatial conflicts or clashes, contributing to quality control, robotic fabrication of components, and progress tracking. These applications take advantage of data generated during the design process, and these construction insights can often improve design if incorporated into the process early on. As a result, traditional contractual relationships within the industry are under pressure to change.

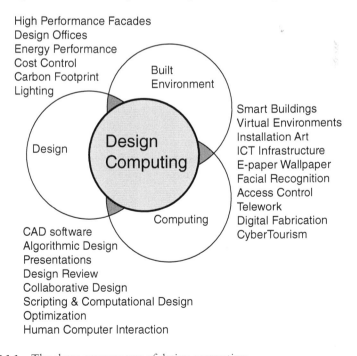

FIGURE 1.1 The three components of design computing.

Less obviously, computing is one component a designer may use in fashioning a smart built environment, where it can contribute to many building systems, influencing shape, security, and energy use. Simulation tools forecast how a design will behave, but sophisticated sensors and building operations software also influence the user experience of the building's real-time operation. Finally, as occupants of built environments and consumers of digital media, designers themselves are influenced by the changes the technology has already wrought and is in the process of delivering.

The emerging discipline of design computing is very broad, spanning the very wide set of interests outlined above, including overlaps or influences in each pair of concerns from Figure 1.1—influences that often flow in more than one direction. The investigation begins with a quick look at each of them.

The Built Environment

The built environment surrounds most of us most of the time. As a product of human culture, this environment is a slow-motion conversation between designers, builders, and users, involving designed objects from the scale of the doorknob to that of the city. This conversation is made up of personal manifestos, esoteric debate about style and importance, and public policy debate about how we should conduct our lives. Increasingly, these utterances are enabled and influenced by digital technologies used in design and construction. Digital tools are increasingly important, if not central, to both processes.

In recent years, in response to population growth and climate change, public discourse has focused attention on energy and material resources used in buildings. In 2014, according to the US Energy Information Administration, 41 percent of total US energy consumption went to residential and commercial buildings (USEIA 2015). Burning coal produces 44 percent of that energy (UCS 2016). The CO_2 produced by that combustion is thought to be a significant contributor to climate change. These facts are of interest to the populace in general, as well as to designers, who must respond professionally to this situation. In fact, beginning in the 1970s, concerns about energy efficiency in buildings led to the development of sophisticated light- and energy-simulation software. These tools, in turn, have resulted in more simulation-based estimates of design performance and justification of design choices, in place of older, less precise prescriptive approaches.

At the same time, the built environment is also the frame within which most of our social and work life is played out. Its organization is widely recognized to influence social and economic behavior. Industrial activity, heavily dependent on movement of goods and materials, originally gave rise to time and motion studies and ideas of optimal layout. After World War II the idea that procedures might be more important than individual designers in determining design quality fostered the *design methods* movement, which was much more sympathetic to applying computational approaches to design. As computers have become more available to design researchers, techniques have grown up around complex computational or statistical approaches. *Space Syntax* researchers (Hillier 1984) have attempted to identify those qualities of the environment around us that delight, distract, or otherwise influence us as we move about the city or carry out work in an office setting (Peponis *et al.* 2007).

New technologies, from elevators to cell phones, have had both minor and profound impacts on what people do in a building, how the designers and builders work to deliver it, and how it impacts society at large. Computing plays a two-sided role, as both a tool for designers to work with (e.g., by enabling remote collaboration) and an influencer of that work. Further, as information and computing technologies (ICT) become less and less expensive, it has become possible to utilize them in different ways. Coffee shops serve as casual offices for many workers because of widespread broadband connectivity and cheap wireless access points. Many houses now include a home office to facilitate telecommuting. Smart buildings may soon recognize us, unlocking our office doors and signing us into the building as we approach. Mechanical systems already take into account weather forecasts and schedules to pre-cool or -warm meeting rooms. In the individual office, email and phone calls might be automatically held during periods of intense concentration or during meetings. Advanced packaging or marketing promises to make any surface in your environment an interactive order point for purchasing (Amazon 2015), but also for information display. It's a very short step from there to the refrigerator that knows how old the leftovers are and when to order milk. Internet-connected toasters, thermostats, door locks, garage-door openers, energy meters, sprinklers, and lights are just part of the emerging "Internet of Things" (IoT). Provision of supporting infrastructure and selection of system elements for the IoT is very likely to fall into the space between occupant and builder—the space of the architect.

Design

> It is widely held that design consists of an iterative cycle of activities in which a design idea is formulated in response to requirements and tested for sufficiency to those requirements. These activities are typically defined as *analysis*, in which the design requirements are formulated from a study of a problematic situation, *synthesis*, in which alternative solutions to the problem are conceived and documented, and *evaluation*, in which the solutions are tested for predicted performance and judged for suitability and optimality.
>
> Clayton et al. 1994, citing Asimow, 1962

Investigation into how best to apply computing to design began in the 1960s, but attention to design process goes back much further. You probably already have at least a passing acquaintance with design process or products. You may have heard the Goethe quote about "frozen music" or the 2000-year-old Roman architect Vitruvius' assertion that architecture's goals are "commodity, firmness and delight" (Vitruvius

15 BC). Importantly, a distinction is often made between buildings with these features and what John Ruskin called "mere building" (Ruskin 1863). These aesthetic and poetic descriptions challenge us to deliver our best design, but they do not provide much of an operational or scientific understanding of design. To complicate things a bit more, architects are only licensed and legally charged to protect public safety in the process of modifying the built environment. Such concerns may seem more appropriate to "mere building," but even those more prosaic challenges become substantial as systems for occupant comfort and safety grow more complex and consumption of materials, power, and water become more problematic.

The challenge of establishing a procedural model of design, while slippery, has been the subject of much attention. Nobel laureate economist and artificial intelligence visionary Herbert Simon, in his book *The Sciences of the Artificial*, argues that we should carefully study *made* things (the artificial) as well as natural things, and goes so far as to outline a curriculum for a "Science of Design." Looking at design through its practice and pedagogy, philosopher and professional education theorist Donald Schön characterizes designers, along with doctors and musicians, as *reflective practitioners*, educated through a mix of mentoring and task immersion (Schön 1984). And, of course, while the nature of design remains a sometimes-contested subject, designers often exhibit observable behaviors that most professional designers and educators would recognize.

For example, architects often study existing environments and they may travel extensively to do so, at which times they tend to carry cameras and sketchbooks to record their experiences. They also study the behavior of people; they meet with users and owners of proposed buildings; they sketch, they observe, they paint; they draft detailed and carefully scaled plans, sections, and elevations as legal documents; they select colors, materials, and systems; they persuade community groups, bankers, and government agencies; and they consult with a variety of related specialists.

Armed with such observations, academics, software developers, and architects have developed many digital design tools by creating one-to-one digital replacements or assistants for the observed behaviors. While this has given us efficient tools such as word processors and CAD drafting, more revolutionary impacts might be obtained by understanding the complete package in terms of its goals, cognitive frames, frustrations and satisfactions. Nonetheless, the word processors and spreadsheets produced by this approach now go so far beyond their antecedents that they have significantly disrupted and reconfigured daily office work and power. Digital design tools seem destined to do the same to design processes and offices.

Design Processes

There is little debate that there are known processes for making design happen. They may not be the same for each designer, but they often include some or all of the stages shown in Figure 1.2, moving from a problem statement to analysis of the problem, generation of one or more ideas, representation of the ideas in a suitable

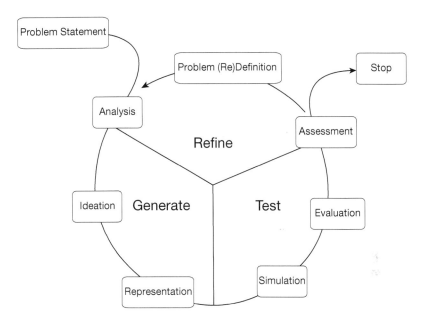

FIGURE 1.2 A simplified model of the design cycle.

medium, simulation of visual and/or other characteristics, evaluation of those results, assessment of the overall design, after which another cycle of refinement or redefinition carries forward until a suitable design is produced and the cycle stops. Design combines elements of both art and science, of spontaneous creativity and labored interaction with the limitations of the real world. While descriptions of art frequently reference inspiration, novelty, creativity, thinking outside the box, divergent thinking, or problem solving through problem subdivision and resolution, design is perhaps best characterized as an iterative process employing cycles of divergent and convergent thinking, aimed at solving a problem through both generation and refinement of ideas. In this process the designer analyzes the problem, establishes constraints and objectives, generates or synthesizes one or more possible solutions that meet some or all of the constraints and objectives, subjects the candidate solutions to assessment, and evaluates the results. The evaluation is used to refine the problem definition, help generate another possible solution, and so on. At some point the evaluation (or the calendar!) suggests that further change is not required or possible and the process stops.

The description above invokes a number of concepts that connect to computing, including representation, transformation, collaboration, solution spaces, generative processes, simulation, evaluation, and optimization. Software exists to do these things with varying levels of ease and success, and more is appearing all the time. Designers may use the software because we have to or because it helps to do the job better. Our practices and professions may be changed by it, as we may individually benefit from it or be replaced by it, and we may out- or underperform in comparison. Designers may also contribute to the development of new software, utilizing insights from practice, facilitating future practice. Finally, we may deploy software and hardware as part of our designs. To think further about these potentials we need to develop some computing concepts.

Computing

Humans have been interested in mechanical computation for centuries, but as the industrial era got rolling in the nineteenth century, Charles Babbage took on the challenge of actually producing an entirely mechanical computer, which he called the *Analytical Engine*. While not entirely successful in his lifetime, his designs have been realized in recent years as operational hardware. Babbage's mechanical calculators are now seen as precursors to the electronic computers of the twentieth century, and his collaborator, Ada Lovelace, is often cited as the first programmer because of her insights into the possibilities of computation.

The Roles of Representation and Algorithm

Babbage's names for the two main parts of his mechanical computing machine were "Store" and "Mill," terms that still nicely capture the two main elements of computer systems, but which we now call random access memory (RAM) and central processing unit (CPU). The Store held numbers needed for subsequent calculations as well as finished results. The Mill was where the computational work was done. Information was shifted from Store to Mill, where it was processed, and then shifted back into storage. Since the Mill was capable of performing several different operations on the values from the Store, a list of instructions specifying which operations to perform and in what order was required as well. Today we would recognize that as a program.

There is no value in storing information that you aren't going to use. But the information required to solve a problem may not be immediately obvious. In fact, the particular information chosen or used to "capture the essence" of a problem therefore *models* or *represents* the problem. It may seem that the appropriate representation is obvious, and it often is, but there are subtle interactions between cognition, representation, problems, and the programs that manipulate the representation.

The programs that control the Mill, or CPU, perform sequences of computations. Different sequences may be possible, and may produce different results or take different amounts of time. Certain commonly occurring and identifiable sequences of computational activity (sorting a list of numbers, for example, or computing the trigonometric sine of an angle) are called *algorithms*. The computations are both enabled and limited by the information stored in the representation, so there is a relationship between the representation used and the algorithms applied to it that also bears examination.

Of course, given an existing representation, if the software allows, we may be able to create a new set of operations to perform in order to produce new results. The contemporary interest in scripting in the field of architecture arises from this opportunity to manipulate geometry and related data through algorithms of our own devising, giving rise to terms such as *computational design*, *emergent form*, and *form finding*.

These relationships are illustrated in Figure 1.3. At the center are the interrelated data model and algorithms around which the application is built. The diagram also reminds us that the information the computer works with is not what exists in the real world, but information filtered by input hardware and software. Continuous things (time, shapes, sound) are sliced into pieces. Numbers representing sound, position, color, and even time all have finite range. There is finite storage and processing power. Mice and cameras record information to some finite pixel or step size. Commands—inputs that trigger change or output—too, are drawn from a finite palette of options. At the interface we also see adjustments to the real world—jitter taken out of video, parallax compensation on touch screens, hand-tremor and limited zoom options eliminated with object snaps and abstractions like straight line primitives, characters constrained by fonts and encoding schemes, etc. New information (e.g., action time-stamps, identity of user) may be added to the data at the same time. During output back into the real world, systems exhibit limitations as well. Displays have limited size and brightness for showing rendered scenes; they use pixels which adds "jaggies" to diagonal lines; output color and sound range are limited by the inks and hardware used; digital sound necessarily builds waveforms from square waves; step size on CNC routers limits the smoothness of surfaces, etc. While we overlook and accept many of these unintended artifacts, we ought not to forget that they occur.

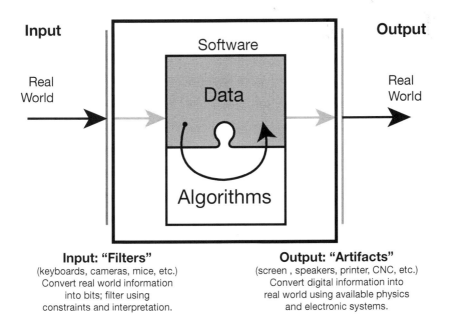

FIGURE 1.3 The relationships between software as data + algorithm and the real world.

Symbol-Manipulating Machines

A century after Babbage, Alan Turing, World War II code-breaker and one of the founding fathers of computer science, described the more abstract symbol-processing nature of these machines and defined a hypothetical test for machine intelligence called the "Turing Test." As with a nuts-bolts-gears-and-pulleys machine, a Turing machine is rigorously predictable in its basic operations, but like any very complex mechanism, it isn't always apparent what the behavior will be after casual (or even careful) inspection. It is a machine, but results can still be surprising, and surprisingly useful, even smart. Turing knew this as well; his Turing Test, which defines *artificial intelligence* (AI) in terms of our ability to tell a human and a computer apart through communicative interactions, remains one of the standard litmus tests of AI (Turing 1950).

Virtual Machines

Computers are both familiar and mysterious for most of us—often used daily but understood only at a superficial level. In their inert, unpowered state they are just simple rectangular objects with a glass panel and buttons on them. But when powered-up, modern computers are chameleons, able to shift rapidly from role to role: typewriter, book-keeping ledger, artist's canvas, designer's drawing board, movie screen, or video-phone. This is often called *virtuality*—the quality of "acting like" something well enough that you don't notice the difference. (Had this usage been popular sooner, the Turing Test for AI might be called a test of virtual intelligence instead). Virtuality arises from the computer's ability to store and recall different programs, combined with its ability to present different visual environments on the screen, and our willingness to cognitively remap the (increasingly rare) "virtual" experience to the "non-virtual" one. Virtuality arises from both hardware and software features—our ability to capture, or model, interesting activities in digital form well enough to elicit a suspension of disbelief. We have been so successful at this that some familiar activities, such as word processing, spreadsheet manipulation, and web-browsing may seem inconceivable without a computer.

Design Computing

The application of computing to design began with high aspirations, but ultimately it was found that it was easier to progress by harvesting the low-hanging fruit, making one-to-one substitutions for existing tools. Beginning before the widespread access to graphic displays, the focus was on descriptive and analytical applications, not drafting applications. One-to-one replacements appeared for the most straightforward activities, including core business practices such as accounting and text-based communication in the form of spreadsheets and word processing. Research was focused on computationally heavy activities such as thermal and structural analysis, layout optimization, and specifications. While there were few interactive graphics displays, there were plotters, and plan drawings, wireframe perspectives and even animation frames produced from decks of punch cards. Interactive raster displays brought big changes.

For designers, one of the watershed moments in the history of computing came in the mid-1980s, when computers began to produce graphics. Until then computers were largely about text and numbers, not graphics, pictures, or drawings. While a graphic display was part of Ivan Sutherland's 1963 *Sketchpad* research, and both vector-refresh and storage tube displays were available throughout the 1970s, graphic displays didn't become widely available until the early 1980s. Only with the advent of the Apple Macintosh, in 1984, was the modern graphical user interface (GUI) firmly established, and the full potential of computer-as-simulacrum made visible.

Once interactive graphics appeared, much more attention was paid to drawing or drafting, to the extent that most readers may think about "computer aided design" (CAD) only in terms of drafting software. Several factors contributed to this situation in the 1980s, including the fact that 40 percent of design fees were paid for production of construction documents, two-dimensional CAD software was less demanding on hardware, and it was easy for firm management to understand and implement CAD (i.e., document production) workflows since CAD layers closely mimicked the pin-drafting paradigm in use at the time and much of the work was done by production drafters working from red-lined markups.

We can build a broader understanding of the computational approach to design by paying careful consideration to the other steps involved in the overall description of design and to the traditional strategies employed to address those challenges. These tend to include almost equal amounts of the artistic (creative ideation) and the analytical (numerical assessment). As we will see, purely computational solutions are elusive, so one of the great challenges of design computing is to find the appropriate distribution of responsibility in order to balance human and computer activities in the design process.

Computers can either facilitate or hinder design activities, depending on how they are used within the design process. While we often hear about "virtual" this and that, the truth is that computers are used to change the real world, even if that just means making us feel good about our video-gaming prowess. To accomplish this, they must transform the world (or some part of it) into a digital representation. This process depends on the features and opportunities, or *affordances*, of the input hardware (Gibson 1986), as well as the requirements of the internal digital model or representation, combined with the software affordances desired. Information will be added or lost in the process (for example, our imprecise hand motions might be corrected to produce perfectly horizontal or curved lines, or our image might be pixelated by our webcam's image sensor). We might then transform that digital model in small or big ways before outputting it to the real world again, at which point the mechanical and electrical affordances of the output device (screen size, color-space of the printer, router speed or step-size, printer resolution, pen type or color in a plotter, etc.) further influence the result. By learning about these influences, we become computer literate. If we know about the internal representation and related transformations, we can take charge of our workflow.

Design Spaces

Complex representations are built up out of simpler elemental parts, in much the same way that your online gaming profile (which represents you) might contain a screen-name (text), age (number), and avatar picture (raster image). If our goal is to define a unique gameplay identity, it must be possible to represent it, giving it a value for each of the elemental parts. If you think of these as axes in a coordinate system, you can think of each profile as a (name, age, avatar) "coordinate." The game's profile representation can be said to offer a "space" of possible identities. The problem of creating, or designing, our unique identity requires "searching the design space" for a suitable representation that hasn't already been claimed.

Architectural design problems involve a great many elemental components, many of which (e.g., room dimensions) can take on a very wide range of values, creating very large—even *vast*—design spaces (Woodbury and Burrow 2006). This means there is ample opportunity for creativity, but it complicates the question of automated computation of designs. Unfortunately, really creative human solutions sometimes involve defining entirely new representations, which is something of a problem for computers.

Interoperability: Data and Programs

Partly because of virtuality, we have to tell the computer what "tool" to be before we do something with it. That means we control what the machine does by selecting *both* the information we wish to process (the *data*) and the type of process (the *program*) we wish to utilize. The two parts fit together like a lock and key.

As a consequence, the data used by one program will often be useless for another, even when they might represent the same building. This characteristic has broken down somewhat in recent years, as developers have struggled to make interoperable representations, but it still remains largely the case, and it may be fundamental to the nature of computing, since alternative algorithms may require different representations and knowledge about that data. This tends to make data proprietary to the program with which it was created. It also makes it difficult to dramatically change the representation without breaking the program.

Exceptions to this general rule occur when data formats are documented and the documents made publicly available so that programs can be written to "Save As…" to the desired format, or when programs are written to accept data in a pre-existing public format, as when Industry Foundation Classes (IFC) or Initial Graphic Exchange Standard (IGES) is used to import CAD data into a construction scheduling program. Still, programmers acting with the best of intentions make alternative interpretations of documents, which can lead to incompatible implementations and frustrating responses by inflexible machines.

Computational Complexity

Even our super-fast modern computers require some time to compute a sum or compare two numbers, with the result that complex computations can take perceptible amounts of time. Since each operation requires a different amount of time to perform, each algorithm can be scored according to the time it takes. For example, during playback your MP3 player has to read the music data off the disk and perform the appropriate computations to play the next moment of music before the last one has died away. If it takes longer to compute than it does to play, there will be "drop outs" in the audio playback, or "rebuffering" in your streaming video experience.

The representation also has an impact on the time it takes to do something. Storing more data may require bigger files but reduce the complexity of the computations required to play them. Alternatively, more computation done as the music is being recorded may save file space and make the playback computations straightforward.

We can characterize many problems and their solution algorithms according to their complexity. For example, imagine you have to visit each of five cities and wish to find the lowest-cost driving route. One way to do this—one algorithm—is to compute the number of miles required for every possible route and select the route with the smallest value. Sounds simple enough, right? To investigate the complexity of this situation, you need to consider the number of possible routes. If there are five cities on our list then there are five possible "first cities." Once a city has been visited, we don't go back, so there are four possible "second cities." That makes 20 combinations for cities 1 and 2. There are three "third" cities for a total of 60 combinations for 1, 2, and 3. Selecting one of the two remaining cities makes for 120 combinations. When we choose between the last two cities we indirectly choose the very last one, so the total is $120 = 5 \cdot 4 \cdot 3 \cdot 2 \cdot 1$—five factorial (written as "5!")—combinations. While other factors will contribute to the time required, the total time to compute the shortest route will depend on this "n-factorial" number. As long as we stick with the basic algorithm of "test all possible routes and select the shortest" we have to generate every route in order to test it, and n-factorial grows large very rapidly. For 20 cities it is 2.4×10^{18}. Even computing 10,000 routes/second, you would need 7.7 million years to crunch the numbers. Ouch.

This problem is called the *Traveling Salesman Problem* and is one of the classics of computer science. The "test every route" solution algorithm is easy to describe and relies on simple computations, but due to the *combinatorial explosion* the total computation time required for realistic problems renders the approach useless. Unfortunately, in many design problems, as in the Traveling Salesman Problem, designers seek the "best" combination from among a very large number of alternatives. Seeking *better* rather than *best* is a more achievable goal, but it raises two questions: How do we improve a given design? And how do we know when to stop?

FIT, Optimization, and Refinement

The task of producing better designs, called *optimization*, requires that we solve two big problems: picking design moves and assessing design quality.

In calculus we learn that functions have minima and maxima that can be found by differentiating the equation and solving for the derivative's zero values. Given a mathematical expression for the value of something, a direct mathematical, or *analytic*, means would exist for arriving at the best solution. We can, for example, compute the best depth for a beam carrying a given load. This doesn't usually work for design because there is rarely a clear mathematical relationship between the changes you can make and the performance measures you wish to improve. If a cost estimate is too high, for example, you can't simply make a smaller house—humans, furniture and appliances don't come in continuously variable sizes. Instead, design involves trade-offs: Do you want more, smaller rooms or fewer, larger rooms? Smaller rooms or less expensive finishes?

Choosing the next move is not easy, nor is assessing its impact on the many dimensions of the design. Design quality is often described in terms of the "fit" or "misfit" between problem and proposed solution (Alexander 1964). Where it can be expressed as a numeric value (cost, energy use, volume, etc.), it provides a means of comparing alternatives, sometimes called a *fitness*, *objective*, or *cost* function, but mostly in narrow domains.

Lacking a calculus of design and condemned to incremental improvement searches, machines share with human designers the problem of needing a "stopping rule" that indicates when the design is "good enough."

Fabrication, Automated Construction, and Mass Customization

Prior to the Industrial Revolution, products such as cookware, furniture, and housing were made by hand, and therefore pretty much unique. The Industrial Revolution ushered in the idea of factory-made objects using interchangeable parts. It also introduced the concepts of off-the-shelf or off-the-rack. By standardizing fabrication we benefit from economies of manufacturing scale that make devices such as tablet computers and smartphones affordable by most, but impossibly costly as single-unit hand-made products. In the building industry we use standardized bricks, blocks, studs, steel sections, HVAC units, and tract housing. Now, as fabrication processes become both more robotic and more digital, the cost of custom products, like a single page from your laser printer, is declining, delivering mass customization—economical, personal, unique production—at the level of buildings. While liberating designers to change the rules of design and production, these changes also force designers to take greater responsibility for their work in entirely new ways.

Summary

We have entered a period of rapid change in the character, design, and production of many products, including buildings. Many challenges will be encountered, wrestled with, and perhaps overcome. Designers need to develop or reinforce their understanding of what goes on both inside and outside the black boxes of *design* and *computation*. In Chapters 2–5 we will explore the state-of-the-art in design and computing in some depth. Then, in the remaining chapters, we will consider the fundamental questions and challenges that shape and motivate the emerging field of design computing.

Suggested Reading

Alexander, Christopher. 1964. *Notes on the synthesis of form*. Cambridge, MA: Harvard University Press.

Schön, Donald. 1984. *The reflective practitioner: How professionals think in action*. New York: Basic Books.

References

Alexander, Christopher. 1964. *Notes on the synthesis of form*. Cambridge, MA: Harvard University Press.

Amazon. 2015. Amazon dash button. Amazon.com, Inc. www.amazon.com/oc/dash-button

Asimow, M. 1962. *Introduction to design*. Upper Saddle River, NJ: Prentice-Hall.

Clayton, Mark, John C. Kunz, Martin A. Fischer, and Paul Teicholz. 1994. First drawings, then semantics. *Proceedings of ACADIA 1994*, 13–26.

Gibson, James J. 1986. The theory of affordances, in *The Ecological Approach to Visual Perception,* 127–146. Hillsdale, NJ: Lawrence Erlbaum Associates.

Hillier, Bill and J. Hanson. 1984. *The social logic of space*. New York, NY: Cambridge University Press.

Oxman, Rivka and Robert Oxman. 2014. *Theories of the digital in architecture*. London: Routledge.

Peponis, John, Sonit Bafna, Ritu Bajaj, Joyce Bromberg, Christine Congdon, Mahbub Rashid, Susan Warmels, Yan Zhang, and Craig Zimring. 2007. Designing space to support knowledge work. *Environment and Behavior* 39 (6): 815–840.

Ruskin, John. 1863. *Selections from the writings of John Ruskin: With a portrait*. London: Smith, Elder & Co., 184.

Schön, Donald. 1984. *The reflective practitioner: How professionals think in action*. New York: Basic Books.

Turing, Alan. 1950. Computing machinery and intelligence. *Mind* 49: 433–460.

UCS. 2016. Coal generates 44% of our electricity, and is the single biggest air polluter in the Union of Concerned Scientists. www.ucsusa.org/clean_energy/coalvswind/c01.html#.VWTM0eu24nQ.

USEIA. 2015. How much energy is consumed in residential and commercial buildings in the United States? US Energy Information Administration. http://eia.gov/tools/faqs/faq.cfm?id=86&t=1.

Woodbury, Robert and Andrew L. Burrow. 2006. Whither design space? *AIE EDAM: Artificial Intelligence for Engineering Design, Analysis, and Manufacturing* 20: 63–82.

PART I
Starting Points

Some portion of the subject matter of this book—perhaps all of it—may be unfamiliar to you, or you may be an experienced graphic or industrial designer, an experienced coder, or a student of architecture, but just beginning to grapple with some of the questions raised here. In any case, there are some general understandings of design, the built environment, and computing systems, and concepts that are needed to establish a foundation for further inquiry. The first part of the book presents the most important of these. If they cover familiar ground, they might be skipped or skimmed.

Models. Humans build many different kinds of models, including scale models of buildings, conceptual models of natural systems, and digital models of real-world phenomena that are fundamental to applying the power of computers to solving problems. An examination of some of the characteristics of models is critical to beginning to build a model of design computing.

The Built Environment. Design operates across a great many scales, from graphic design and product design up to buildings, civic infrastructure, and regional planning. Distinct from the natural world, the built world involves human intention and action. The disciplines of urban planning, urban design, landscape architecture, and architecture share many concepts and challenges, as well as many tools. This chapter presents the broad challenges and opportunities of architecture and design as seen through the prism of computing.

Computers. While often presented in anthropomorphic terms, computers are fundamentally complex machines that thrive on predictability. To really understand design computing, we need to become familiar with some of these fundamentals. We need to learn to "think inside the box."

Monkey See, Monkey Do. Application of computer technology requires understanding of both the technology and the application domain. Often, with new technologies, the first phase of application is to mimic pre-existing patterns.

We examine the state-of-the-art with regard to computing tools and design, most of which is based on attempting to recreate in digital form the processes, tools, and representations that existed in the pre-digital, or analog, world. In this we observe the result of approaching design computing by "doing what designers do."

2
MODELS

Computer programmers and architects both build models. While they are quite different from one another in terms of the media they use, they do share some features. We usually think of architectural models as miniature scaled-down physical objects, built to show the overall geometry and sometimes the colors of a proposed building. The models that programmers build describe relationships of cause and effect within a software system. Both kinds of models are simplified or abstract representations of information and processes, simpler than the reality they mean to capture, focused on the essential aspects of the situation, used to enable control or make predictions, because they behave correctly in the limited reality they represent. For instance, an architectural study model is often proportionally correct, but materially inaccurate (few buildings are built of cardboard, balsa, or basswood), and while it may have the correct exterior geometry it often has no interior partitions.

Models dispose of unnecessary complexity, and in doing so they enable us to ask "what if" questions in economical, reversible ways. They offer us a substitute for full-sized, full-cost, and confusing experiences.

Models don't have to be true in any objective sense, as long as they are useful. Models can be used to focus attention; architects build white pasteboard models as a way to study form, intentionally suppressing color considerations. Models can be conventional; consider the persistence of the words *sunrise* and *sunset* to describe a particular relationship of planet, sun, and observer, in spite of the fact that we know the sun is not moving as the words suggest. Models may be created simply because they are convenient. Such is the case with "Hem-Fir," a fictional tree species for which the Western Wood Products Association publishes allowable stress values (WWPA 2016). A close reading of the description reveals that each individual board in a bundle of Hem-Fir delivered to a job-site is either hemlock or some true fir, such as Douglas fir. The two species of tree grow together and have similar appearance and strength values, so separating them during the harvesting and milling process seems an unnecessary expense. Since most of this lumber is used in stud walls and joists, repetitive-use situations, a statistical abstraction called "Hem-Fir" was created to describe it. The species is a fiction—a model—but a useful one.

Symbolic Models

Drawing on laws of physics, and assuming uniform material properties, uniform dimensions, and uniform temperatures (called *steady state*), it is often possible to derive algebraic equations to describe aspects of architecture, such as the thermal performance of a wall, or compute the shear and moment stresses in a beam. These are *symbolic models*. Their results can be computed directly when the equations are combined with empirically measured material properties.

Finite Element Models

Symbolic models break down when the steady-state assumptions underpinning them are violated—that is, when they are subjected to dynamic (time-varying) influences (wind, earthquake, and impact for structures; diurnal temperature changes, or solar irradiation for thermal). Under such conditions the basic rules of physics still apply, but since it takes time for change to propagate through the structure, the simple steady-state equations don't produce correct results. To work around this difficulty we slice time (and space) into smaller pieces, returning them to a near-steady-state situation, using *finite element analysis* (FEA) to create a *finite element model*. Computing results from the model requires lots of computation for all those small slices of time and space, but computers are good at that.

In the case of thermal behavior, the usual approach to this problem is to divide the building into a network of individual elements, called nodes, connected together by their ability to store and exchange heat. Included in this network are the interior air volume, the outside environment, the ground, the various wall materials, and so on. At any given moment of time the laws of physics apply, so if we divide time into short increments (perhaps an hour instead of a year or day), we can compute how much heat flows from node to node. An hour later there may be different weather, a different thermostat setting, and a change in the number of occupants. Once those changes are taken into account, we can once again compute heat flows in the network of nodes. Repeating this cycle, we can project, or simulate, how much energy will be used by the building over a period of time.

Models like this, consisting of a finite number of nodes and a finite number of time slices, are called *finite element models*. The technique is approximate, in both space and time. More nodes or more slices will produce a marginally better result at the expense of larger data files and more time spent computing the additional interactions. As with other models, making the right simplification is an important part of getting good results; the art of finite element modeling resides in knowing what the appropriate level of abstraction is for the problem at hand. Walls in shade and walls in sun are different, as are north- and east-facing walls, or walls around the garage versus walls around the occupied parts of the house. Distinguishing among those elements makes sense, but dividing the amount of east-facing wall that is sunlit into more parts just to have more nodes does not. Similarly, since simulations such as this usually rely on measured weather data from a nearby weather station, and such measurements rarely occur more often than hourly, it may not make sense to divide time more finely.

Statistical Models

Simulating energy use for a city using a finite element model of each building would be cumbersome and extremely slow due to the large number of variables. However, there is probably a supply of historical weather and consumption data, as well as property tax data about construction and size. By extracting data from similar buildings in similar circumstances, we can build a *statistical model* of the built environment under study. Both real-estate appraisers and tax appraisers use such techniques to set a theoretical sale price on your home, while contractors use it to estimate the cost of doing a kitchen remodel. Estimates of bus ridership, traffic density, pollution, and weather all involve statistical models of the phenomena in question.

Analogue Models

As it happens, the equations for material resistance to heat flow and storage turn out to be mathematically the same as those for electrical resistance and capacitance. Using common electrical components—capacitors and resistors—and replacing temperature differences with voltage differences, we can literally build an electrical circuit to simulate the thermal behavior of a building (Robertson and Gross 1958). This is an *analogue model*; it need bear no physical or visual resemblance to the real thing, but its behavior is used to predict, by analogy, how that system will perform.

Some analogue models *do* look like their subjects. Because light behaves the same in a full-size environment as it does in a small one, we can build *scale models* of buildings using materials such as cardboard, colored paper, and fabric that mimic the visual properties of finish materials. When lit with light from an artificial or real sky-dome, the physical model functions as a powerful simulation of the design's visual environment.

Of course, architects routinely build scale models as part of assessing the visual character of designs, but we don't always consider why such models are actually meaningful. Not all physical models scale accurately. For example, because thermal characteristics are proportional to volume and area, a dimensionally scaled physical model using real materials does not perform the same as proportionately identical full-size objects. Thus, models are not all equivalent, and any model may also produce inaccurate results if applied to the wrong situation.

Sources of Error

Finite element simulation of lighting, heat transfer, structures, and air-flow is now routine (Kensek 2014), but such models can produce erroneous results. They depend on simplifications chosen, the duration of the time-slice used in the simulations, the accuracy of material properties assigned to elements, and the accuracy of the applied data—weather, traffic, etc. Further, the model may exclude a phenomenon that is, in fact, linked to the one being studied. This is why heat and light are often simulated together; people need warm spaces and light to work, and lights produce waste heat, so lighting and thermal performance are tightly coupled with occupancy patterns in most buildings and are also linked to building operation strategies.

Selecting appropriate test data can be a challenge. For example, thermal (energy) simulations need weather data to work with, either typical (for average performance) or extreme (to test worst-case scenarios like a heat wave). Historical data is best, but no randomly selected interval is really typical or extreme. Getting, or synthesizing, appropriate weather data is something of a challenge (Degelman 2003). Further, weather found on a real building site may vary fairly markedly from that of the nearest airport, where the local weather data is collected. Localizing weather data and accounting for site conditions such as trees or buildings on adjacent property becomes more important as simulations get more precise.

Overly simplified or misapplied models can be a problem for inexperienced users. Models are, by their nature, simplifications, often explicitly intended to make complex problems easier for non-experts to work with, making them "designer friendly." At the same time, questions that motivate consultation of the model may be simple or complex. Overly complex tools don't fit simple questions, and simple tools may not fit complex situations. The potential always exists for a mismatch or misapplication due to some unexpected characteristic of the building being studied. For example, one sophisticated thermal simulation of a Seattle-area home that included a glazed atrium showed very elevated mid-winter temperatures. The mechanical engineer running the simulation doubted the validity of these results. After increasing the frequency with which the convective heat-transfer coefficient in the atrium was re-computed (a time-consuming calculation that was minimized for normal conditions), results returned to the more typical Seattle range. Experiences such as this lead some researchers to question the whole idea of "designer friendly" simulation tools (Augenbroe 2003).

Even when appropriate tools are paired with informed questions, there is a risk that some of the information going into the simulation will be incorrect. This is true even in an increasingly connected and online work environment. Dennis Neeley, whose companies have been doing electronic data publishing for manufacturers since the 1990s, has stated that "the challenge is getting correct information…. On some projects getting the information right takes longer than any other part of the project" (Neeley, personal communication).

Even when the input data is correct and the algorithm is appropriate, the fact that digital models use approximate representations of some numbers can cause errors to creep into calculations that are repeated thousands or millions of times. These have receded in recent years as larger system memories, better processors, and sophisticated software are applied to the problem, but these "round off" errors do occur (Goldberg 1991).

Another source of error arises from the common, possibly necessary, duplication of files. What might be called "divergent evolution" happens when data is duplicated for any reason, and then edited independently from the original, as when a consultant is given a "background" drawing of a design. Sometimes it is done to "test an idea" or "try something out." At other times it happens because a backup is mistaken for the primary file, or someone needs to work at home over the weekend. It can even happen within a file, as when work is duplicated from one layer to another "temporarily." In CAD systems, merging (or rejecting) the changes is usually visual, time consuming, and error prone. BIM software tends to be better at alerting users to incompatible edits as long as they are working on the same file (which is rarely the case with consultants). In the software and document management world there are tools for finding and highlighting differences between versions. More development is needed in the world of geometry and design.

Summary

Models are important, there are several types, and they need to fit the problem. Choosing the right model and matching it to the level of detail available in the design data and to the expectations of the designer is important, and there is much room for improvement. While we need to know the limits of our models, the parts of reality they accurately predict and the parts they do not, being computer code they are often opaque and under-documented or poorly understood. When we make an intercontinental phone call we need to remember that sunset in London does not happen at the same time as sunset in Seattle, and when we pick up a Hem-Fir board we should not be surprised that it is one or the other. With digital models, opaque to casual observation, it is harder to discover the limits and harder to correct errors, an observation that has motivated a move to develop tools with more *transparent interfaces* that allow users to see into internal software processes (Tanimoto 2004). In the absence of such software, knowledgeable users remain the primary defense against errors.

References

Augenbroe, Godfried. 2003. Trends in building simulation, in *Advanced building simulation*. Edited by A. Malkawi and G. Augenbroe, 4–24. New York, NY: Spon.
Degelman, Larry. 2003. Simulation and uncertainty: Weather predictions, in *Advanced building simulation*. Edited by A. Malkawi and G. Augenbroe, 60–86. New York, NY: Spon.
Goldberg, David. 1991. What every computer scientist should know about floating-point arithmetic. *Computing Surveys* (March): 5–48.

Kensek, Karen. 2014. Analytical BIM: BIM fragments, domain gaps and other impediments, in *Building information modeling: BIM in current and future practice*. Edited by K. Kensek and D. Noble, 157–172. Hoboken, NJ: Wiley.

Robertson, A. F. and Daniel Gross. 1958. Electrical-analog method for transient heat-flow analysis. *Journal of Research of the National Bureau of Standards* 61 (2): 105–115.

Tanimoto, S. L. 2004. Transparent interfaces: Models and methods. *Proceedings of the AVI 2004 Workshop on Invisible and Transparent Interfaces*, Gallipoli, Italy. New York, NY: ACM.

WWPA. 2016. Framing lumber: Base values for western dimension lumber. Western Wood Products Association. www.wwpa.org/Portals/9/docs/pdf/dvalues.pdf.

3

THE BUILT ENVIRONMENT

> Une maison est une machine-à-habiter. (A house is a machine for living in.)
> *Le Corbusier, Vers une architecture [Towards an Architecture] (1923)*

The architecture, engineering, and construction (AEC) industry has encountered a number of challenges and opportunities in recent years. There is increasing attention to the way that construction and operation of the built environment uses resources, significant change to the technology of design and construction through the application of computers, and broad changes to the technology deployed within environments by occupants and operators. Spanning both high and low culture—from signature buildings to prosaic warehouses, designers and contractors stand out as having a high impact on our quality of life and are seen to operate in a rapidly evolving technological domain. Yet there seems to be little industry-wide benefit derived from the increasing use of computing.

Curators of the Environment

In the developed world, the impact of the built environment is increasingly recognized. A. H. Maslow's hierarchy of human needs identifies five levels: physiological, safety, love/belonging, esteem, and self-actualization (Maslow 1943). At each level our ability to pursue those needs depends on satisfaction of the lower-level needs. The built environment spans Maslow's hierarchy from the bottom to the top, from basic shelter, bedrooms, kitchens, and workplaces to social halls, civic spaces, and environments of religious or creative expression. Building can fulfill a client's needs for esteem and self-actualization. Designers themselves may find such satisfaction through projects, but we also recognize that each project touches the lower levels of the hierarchy and other people's needs, using resources or creating environments that affect many members of the public in addition to the client. As a consequence, our governments construct some parts of the environment, such as roads and transit systems, and restrict what we are allowed to do on other parts through building and land-use codes.

Designers have a bidirectional relationship with the built environment; we are both its creators and creations, albeit in different moments of time and space. The common image of the architect is that of the form-giver, but the built environment, while physical, is also a cultural product, and each building contributes something profound or mundane to the conversation about that aspect of our lives. As creatures of our respective cultures we reflect values and norms that may be different from others, even when our vocabulary is not. For example, architecture students in England have a very different internal image of a "vacation lodge" than do architecture students from the western half of the United States. UK students tend to think cozy wood-paneled country pub, while their US counterparts go with something more along the lines of the soaring lobby of Yellowstone Lodge.

A couple of hundred years ago, European students of architecture went to Rome or Greece to learn about architecture, in contrast to Ruskin's "mere building." There was a fairly limited palette of forms from which to draw. The buildings were stone or brick, plumbing was simple, the city layout grand or mundane. Light came through windows during the day, and from candles or gaslight at night if you were lucky. Heat, when available, came from burning wood or coal, often in the room in question; cooling came from opening the windows. Water moved around in buckets, not pipes, and sewage did the same. For the most part, space was built and then people did their best to make use of it. Palladio's villas had symmetry, proportion, and grandeur, but no master-suites, media rooms, or gift-wrapping rooms.

Buildings and design have grown much more complicated. In the last one and a half centuries, commerce has given access to materials from far away as well as close to hand. We have added electric lights of many types, air-conditioning, elevators, refrigerators, communications systems (digital, video, phones, alarms), fire-suppression systems and fire-stair requirements, mechanical systems to distribute both warmed and cooled air or water, and garages for our automobiles. In more recent years we have endeavored to make our use of resources more efficient. Modern buildings are built to leak less heat and air, sensors turn off lights in unused spaces or at unoccupied times, and thermostats follow schedules that attempt to align the energy investment in comfort to our need for it. Some perform rainwater catchment and gray water recycling. And we now recognize that the differently abled among us should be accommodated too.

At the same time that buildings have been growing more complicated as interconnected systems, we have grown more aware of the human and economic impacts of their configuration and operation. Significant research has gone into understanding how humans work and communicate in critical or complex environments such as hospitals, courts, and factories. We have discovered that we can use design to make factories more efficient, make nurses' jobs easier, and make patients heal faster, while saving energy and other resources. We have also added expectations to buildings in the form of various codes—including energy use, permitted activities, structural requirements for wind and earthquake, lighting, and accessibility. Further, these guidelines and restrictions can vary geographically, so the same building program can produce very different buildings on sites in adjacent cities.

In developing designs, architects are expected to access and work with dynamic and varied information about materials, activities, codes, systems, and locations. As the number of choices grows, they also interact in more complex ways, making it increasingly difficult for an individual to obtain or retain the requisite knowledge. As a result, design has grown more collaborative in nature and buildings less like nice caves and more like machines, as evidenced by the Le Corbusier quote that started this chapter.

Designs, of course, are not buildings. For the last century or so, designers have not been builders; the two have been connected by individual contracts with the client, contracts that have made 2D paper drawings and specifications the principal legal document. While different industry structures exist in other parts of the world, the US construction industry is very fragmented, with many small firms both on the design and construction side. Teams of sub-contractors are assembled for each project and must learn to work together.

In sum, the influences on design are complex, and many influences take the form of location-specific information. Information assembled during the design process needs to be shared with many different participants, ranging from consultants and potential contractors to governmental code compliance checkers and bank financiers. Each wants somewhat different data and may get data at different points in the project's development. Finally, most buildings are unique singular prototypes that will be manufactured only once. Overall, there is limited chance to optimize the design and production process.

The Missing Productivity Gains

While each building remains a one-off prototype, they are much more complex, and now require a skilled team to get them designed and built. The design team for even a modest building includes architects, structural engineers, mechanical engineers, electrical engineers, and lighting designers. During construction, the number of trades involved from the building industry might be double that number. In more sizable or challenging projects there may be experts in geotechnical issues, security systems, data and communications, elevators, day-lighting, fire-suppression, hazardous materials, industrial processes of one sort or another, specifications, accessibility, LEED certification, etc. For every one of these specialties in design, there exists a parallel construction specialty, and both design and construction expertise is very likely drawn from a large number of separate small contractors. One market research report in 2015 estimated that

> 64.6% of firms with a payroll have fewer than five employees and only 3.4% of firms have more than 100 workers. The industry's low level of

concentration means that no firm holds a dominant position in the market; IBISWorld estimates that the four largest players account for less than 5.0% of industry revenue in 2015.

(IBISWorld 2015)

In light of the above, it may not be surprising that researchers at Stanford University found that output of the US AEC sector has declined in the second half of the twentieth century, though other sectors of the economy have seen significant productivity growth (Figure 3.1; Teicholz *et al.* 2001). That growth is usually linked to the introduction of information technology. Though the conclusions have been challenged as not properly accounting for the increasing shift from on-site to off-site construction (Eastman and Sacks 2008), a recent reconsideration (Teicholz 2013) reaffirms the basic conclusion that productivity in the AEC industry has declined, while general manufacturing has increased dramatically.

This is just one symptom of increasing dissatisfaction with traditional practice. Increasingly, large government and corporate clients have demanded improved information use and provision by their architects (CURT 2004). For example, clients have historically specified their needs in terms of building activities and areas, but it was difficult to check a design proposal for compliance with these needs. Similarly, while reduced energy use has been of increasing importance in building requirements, it has been very difficult to get accurate assessments of relative energy efficiency of different schemes. The US General Services Administration (GSA) confronted this challenge in the mid-2000s with a general commitment to BIM processes, beginning with a requirement that design proposals include spatial occupancy plans in a form that could be checked automatically. Simultaneously they encouraged vendors to deliver suitable technology in their

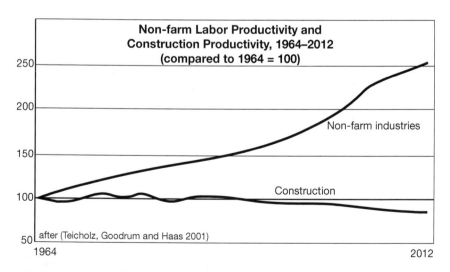

FIGURE 3.1 Motivating the shift to BIM: the "missing" productivity.

software products. As a large and reliable client for buildings of many sizes and in many places, this gave the BIM market a major boost.

There are influences beyond technology and complexity that have slowed AEC productivity gains. The vast majority of the firms involved are small businesses, limiting the amount of vertical integration that can take place. Every building is a one-off project, making it hard to carry over learning from one project to the next. The most common means of communicating information between participants has been the legally binding 2D construction documents. Finally, the design–bid–build procurement process of some entities (e.g., government agencies) often limits the degree to which information can be shared between the designers and contractors, and complicates the financial picture when those who develop the data are not those who benefit financially from its use.

The shift from CAD to BIM technology is just one of the changes in the structure of the AEC industry. The other emerging shift has to do with contractual risk and reward structures. Traditionally, architects and builders sign independent contracts with the owner, though one part of the architect's role may be to act as the owner's representative on the job-site during construction. Furthermore, multiple builders bid on jobs, based solely on the paper documents, and public entities are often required to select the lowest bidder. This sets up a tension between the architect and the builder in which the design information (drawings and specifications) are instruments of control and contention rather than a joint expression of design intent.

The last few years, however, have seen the emergence of alternative legal structures (Salmon 2009). The one most commonly tied to BIM is *integrated project delivery* (IPD), in which a team is assembled at the beginning of the project to share risk and reward in the project by sharing information. In a related approach, design–build firms offer both services to the client in one contract. Institutional clients may pursue a more arms-length relationship to the building, hiring the builder to design, build, operate, and maintain (DBOM) a custom-designed building in which they then lease space.

Each of these legal arrangements determines which exchanges of information are easy or hard, risky or not within the context of construction liability. It is not at all uncommon for contractors to re-do the geometry model of the project, both because they need different information from it and to check that the parts go together "as advertised." This seems wildly inefficient until you recall that design drawings are necessarily incomplete and that various sub-contractors have historically made "shop drawings" of their deliverables for architects to check.

The Internet of Things

At the same time that information technology has presented opportunities and challenges to designers, the opportunities to make creative use of ICT within the fabric of new buildings have also been expanding via the *Internet of Things* (IoT). Increasingly, Le Corbusier's notion of the house as "a machine for living in" has been coming true as our information devices become both more intimately connected to us, and more interconnected with each other. Where power receptacles used to host occasional floor polishers or desk lamps and university lecture halls needed only a few, the modern office may need to power multiple computers, phone chargers, printers, and an occasional desk lamp, while lecture hall seats near power outlets are the very first to go—taken by students in search of a laptop charge. Modern classroom refits usually include video projectors, many more power outlets (sometimes one per seat), and enhanced wireless network service.

At the same time, our environment has become home to a great many sensors, some or all of which may connect to the internet. Traditional smoke and burglar alarms have been augmented with CO_2 detectors, moisture and water detectors, and motion detectors that may check for invaders (motion where it shouldn't be), injured elderly (immobility where it shouldn't be), or simple disuse (turning lights off in empty rooms). If you drive off and leave your garage door open, the house may call you to let you know. When it does, you may be able to use an app on your smartphone to command the door to close. If your sister arrives unannounced while you're at work, you might unlock your front door remotely. Or you might configure the house to turn on the front-porch light and unlock the door as you approach.

All of these examples are part of the emerging IoT. This trend suggests or demands that we rethink some of the basic services, materials and features that architects design into buildings. Where a remotely controlled lock is simply an enhancement to existing technology, individually addressable LED lights may spell the end of light switches, while wireless speakers and phones reduce or eliminate the need for pre-wiring. ICT and the IoT are, increasingly, an integral component of design decision-making, not an applique applied after the fact. Some new homes now come with a central server to run security, data, and entertainment systems.

AEC Data

Even without the IoT, building designs respond to and produce significant amounts of data. Information about site geography and geometry, municipal services, climate, neighborhood, client wishes, available contractor skills, budget, materials, and human behavior—all influence the design directly or indirectly. The disparate nature of these data flows and the integrative character of design often conspire to reduce issues to heuristics (rules of thumb) and reliance on experience. Sorting out useful and appropriate ways to bring data to bear on design problems remains a work in progress.

Getting good data is another challenge. While government agencies and online sources may have and make available digital terrain and other data, the ultimate liability for their use remains the designer's. Vendors and manufacturers make product data available online, but they can't always be counted on to provide accurate information. Firms archive old projects, and often mine them for details to use on new projects, which is a productivity boost. Paradoxically, such file archives can make bad data remarkably hard to expunge. It remains a challenge to integrate knowledge and experience into the institutional memory of a practice or historical record of a building.

Municipal, state, and national governments care about buildings, and exercise influence through various codes. Zoning and building codes determine what can be built, how it can be used, and what features it must have. Local restrictions, including covenants on individual parcels, may also apply. Designs must pass both planning and construction inspections for permits to be issued. While most codes are justified as providing a public good, the codes themselves are protected intellectual property, and while there are national and international building code frameworks, the situation is complicated by changes implemented by state and local governments. Thus, the codes themselves are rarely available online except by subscription, and local variations by different jurisdictions are common. Checking code issues on a large commercial project can be a challenge, and automated checks or feedback are not yet available, though progress is occurring in this area (Digital Alchemy 2016).

The construction documents produced in the design process are meant to define projects well enough to get the job done, but they are invariably and intentionally incomplete. This is not simply because designs are works in progress, resolving and reducing ambiguity with each milestone but carrying a residue. Some ambiguity is left, or added, to provide contractors with variability—allowing them to bid a contract based on personal knowledge or connections. Other ambiguity is present to shift responsibility for the result—construction law directs architects to indicate *what* must be built, but not *how* to build it. Finally, incompleteness recognizes that craftsmen on the job will do the work based on their experience and industry standards. There is such a thing as too much information; drawing sets with too much detail often receive bids that are higher.

Building designs are of interest to clients, consultants, contractors, government agencies, neighbors, bankers, and potential occupants, to name just a few. Each of these participants interprets the set of drawings and specifications produced by the design team in a particular way, extracting a different set of facts from the design documents and then combining those with varying contexts, estimates, and assumptions to construct an enhanced and focused representation that serves their needs. The enhanced data is not incorporated in the central data repository, as it may be critical to commercial advantage, irrelevant to others, or ephemeral.

Even when parties share an interest in the same data (e.g., the building geometry), they may use different software to work with that information. Where paper drawings clearly require reconstruction of geometry, the availability of 3D data on the designer's desktop promises an easier process for both design and analysis. One of the features that drew users to Ecotect in the early 2000s was the availability of integrated lighting and energy analyses. Unfortunately, as reported to the 2014 Construction Research Congress, interoperability remains "one of the most important challenges that hinder the adoption of BIM" (Poirier *et al.* 2014, citing SmartMarket Report 2012). Even when the same software is used, differences in how the information is used make interoperation challenging.

When there are multiple potential partners, vendors, and users in a knowledge domain, and no dominant technology supplier, the usual strategy is to assemble a group of representatives who can sit down and hash out standards for the way the data will be handled and created. In the AEC market, there have been efforts to do this at the drawing (IGES, metafile), product (STEP), and assembly or building (IFC, IAI, buildingSMART) level. The International Alliance for Interoperability (IAI), started in 1996, became buildingSMART in 2008 (buildingSMART 2014). It promotes open standards and nomenclature for interoperable software in the AEC marketplace. Curiously, because standards deal in abstraction and serve broad constituencies, one of the products of their effort has been standards-processing software. Indicative of the depth of the challenge, "most standards processing systems developed today assume that the user has an intimate understanding of the classes of objects referred to in the standard and expect their users to describe their problem in these terms" (Fenves *et al.* 1995). Thus, "After … years of development, it is acknowledged that there is no such thing as fast track in building standardization" (Augenbroe 2003, citing Karlen 1995). Without standardized representations, computation about buildings will remain challenging.

Finally, the construction phase of building procurement has arguably benefited more from application of ICT than the design phase, from clash detection among sub-contractor models, to cost estimating and construction scheduling in the office, to automated sheet-metal fabrication and computer-controlled steel rolling mills during construction. Significant cost savings accrue. The problem with this added efficiency is that the architect, or a design consultant, is likely to have created the value that the contractor realizes, which renders traditional cost–benefit sharing formulas out of date.

Summary

Information technologies are part of significant changes to the AEC industry, changing paradigms of project design and delivery, enabling radically different building design and construction processes, changing traditional legal patterns in the process, and influencing the character and features of the spaces being created. At the same time, the industry has not realized the productivity gains seen in other industries after computerization. Possible causes include industry structure, absence of repetition, and incompatible data models.

Suggested Reading

Augenbroe, Godfried. 2003. Developments in interoperability, in *Advanced building simulation*. Edited by A. Malkawi and G. Augenbroe, 189–216. New York, NY: Spon.
Maslow, A.H. 1943. A theory of human motivation. *Psychological Review* 50: 370–396.

Teicholz, P., P. Goodrum, and C. Haas. 2001. U.S. construction labor productivity trends, 1970–1998. *Journal of Construction Engineering Management* 127 (5): 427–429.

References

Augenbroe, Godfried. 2003. Developments in interoperability, in *Advanced building simulation*. Edited by A. Malkawi and G. Augenbroe, 189–216. New York, NY: Spon.

buildingSMART. 2014. buildingSMART: History. www.buildingsmart.org/about/about-buildingsmart/history

CURT. 2004. Collaboration, integrated information and the project lifecycle in building design, construction and operation, WP-1202. Construction Users' Round Table. August.

Digital Alchemy. 2016. Home page. http://digitalalchemypro.com

Eastman, Charles M. and Rafael Sacks. 2008. Relative productivity in the AEC industries in the United States for on-site and off-site activities. *Journal of Construction Engineering Management* 134 (7): 517–526.

Fenves, S. J., J. H. Garrett, H. Kiliccote, K. H. Law, and K. A. Reed. 1995. Computer representations of design standards and building codes: U.S. perspective. *International Journal of Construction Information Technology* 3 (1): 13–34.

IBISWorld. 2015. Architects in the US: Market research report. May. www.ibisworld.com/industry/default.aspx?indid=1401.

Karlen, Ingvar. 1995. Construction integration: From the past to the present, in *Integrated Construction Information*. Edited by P. Brandon and M. Betts, 135–144. New York, NY: Spon.

Le Corbusier. 1923 [2007]. *Vers une architecture* [*Towards an Architecture*]. Translated by John Goodman. Los Angeles, CA: Getty Research Institute.

Maslow, A.H. 1943. A theory of human motivation. *Psychological Review* 50: 370–396.

Poirier, Erik A., Daniel Forgues, and Sheryl Staub-French. 2014. Dimensions of interoperability in the AEC industry. *Proceedings of Construction Research Congress 2014*, ASCE.

Salmon, James. 2009. The legal revolution in construction: How collaborative agreements, BIM and lean construction methods support integrated project delivery. *Journal of Building Information Modeling* (Spring): 18–19.

SmartMarket Report. 2012. *Building information modeling trends*. New York, NY: McGraw-Hill

Teicholz, P. 2013. Labor-productivity declines in the construction industry: Causes and remedies (another look). *AECbytes Viewpoint* 67.

Teicholz, P., P. Goodrum, and C. Haas. 2001. U.S. construction labor productivity trends, 1970–1998. *Journal of Construction Engineering Management* 127 (5): 427–429.

4
THINKING INSIDE THE BOX

A task completed using a computer will generally involve at least three models: one for the data that describes the task; one for the user actions that can be carried out on that data; and one that establishes the cultural context in which it is carried out. The data model determines what is physically possible through the selection of parts and their relationships, the same way that different vehicles (car, bicycle or truck) emerge from choices of frame, axles, tires, power train, and power source. The data model determines, to a great extent, the affordances of the resulting vehicle (how much it can carry, how you steer it), and those affordances establish an interface between the mechanism and the user. Finally, the user's goals and the vehicle's affordances interact with the cultural context—the rules of the road—to further establish whether and how the task can be accomplished.

We don't have to be a car designer or a mechanic to drive, but we do need to know where the tires go, how to add fuel, and when to lubricate it; we learn how to steer, brake, use a manual transmission, parallel park, and change a tire, as well as how fast to go and how to signal a turn. We understand the fundamental components and their relationships to each other. For cars, that produces a certain mechanical literacy; for computers it is often called "computational thinking" and is the subject of this chapter. Because many users learn to compute in a largely task-oriented mode, even experienced and capable users may have incomplete understanding of the system they use, so the focus here is on the low-level model, the mechanics of the system.

A computer is a general-purpose machine for processing information, but its heart is something very much like a simple calculator—enter a couple of numbers, pick a math operation, and it'll tell you a result. To add a list of numbers written down on a pad of paper, you must combine the basic computations with other information (the list) and actions (reading the list, keeping track of your place in the list, knowing when you're done) in a sequence, forming an *algorithm*. Expressed in code the computer can understand, an algorithm becomes a computer program.

Some points to note about even this simple example: (1) neither the initial list of numbers, nor the pattern of operations ("enter, add, repeat") reside in the calculator itself, they reside on the pad of paper or in the user's memory; (2) the user will need to make decisions along the way (e.g., "What's the next number?" "Are we done yet?"); and (3) if you need the final total later, you had better copy it down onto your piece of paper at the end! In the example, the calculator functions as the central processing unit (CPU). The pad of paper and the steps carried out by the human operator are also part of the system. In a real computer the role of the pad of paper is played by memory or disk storage and the central processor or CPU includes hardware to interpret and follow the instructions that make up the algorithm, retrieving each data item in turn and adding it to the intermediate result. In fact, another name for the CPU is "arithmetic and logical unit."

The other important quality this example illustrates is the duality of data and algorithm. Both are needed, and they are interdependent. By combining data and algorithms, computers do amazing things all by themselves, but most of the time they interact with humans (in fact, our smartphones are more about interaction than computation). Their screens can be used to display words, numbers, photographs, and drawings, and their speakers can play sounds. Keyboards and mice let us input information. They are very reliable, fast, and accurate. They don't get bored or tired. They are great partners but they do not make intuitive leaps or get ideas, though they may juxtapose two pieces of information that cause a human to do so.

Virtuality is Real

The different programs you run on your computer control its information-processing behavior, allowing it to switch from being a web browser or CAD program to being a spreadsheet or video game. This characteristic of *acting like* or *presenting the behavior of* something else is referred to as *virtuality*. It's as if all of our kitchen appliances (stove, fridge, sink) were replaced with one piece of hardware that delivers each of these functions when requested.

The computer itself consists of sources of input (including mice, keyboards, cameras, disk drives, and network interfaces, as well as cell phone radios, accelerometers and temperature sensors), the CPU for manipulating the information, short-term memory and long-term disk storage for the information, and a means of communicating information to the outside world (including a display screen, speakers, printer, and network interfaces). The CPU does the actual computation, depending on the instructions in the programs. Information (data) and instructions (software) are both stored in memory during use, where they are quickly accessible by the CPU. The main memory is fairly expensive and volatile (it forgets everything if power is lost), so it is supplemented with non-volatile secondary storage that does not require electricity to retain stored information.

Until quite recently rotating magnetic surfaces were used for this, using various mechanisms ranging from cheap low-capacity "floppy" disks to fast high-capacity "hard drives." More recently flash memory thumb drives and solid-state disks made from computer chips have begun to take over this role, though they continue to be referred to using the old vocabulary, and play the same role in the machine.

When the machine is first turned on, main memory is empty except for a simple program that is actually built into a small area of memory called read-only memory (ROM). On power-up the hardware automatically accesses and runs the instructions in ROM. In a process called *booting* (derived from the phrase "to pull yourself up by your bootstraps") this program retrieves the operating system software from the secondary storage and loads it into memory for active use. The operating system is the software bureaucracy with which you interact and against which all your other applications/programs execute.

When you start up (launch) a program the operating system copies (loads) its instructions from the secondary storage into an unused area of memory and the machine begins executing those instructions. The program may, in turn, copy a data file into other unused memory (opening and reading the file), connect to another computer over a network in order to retrieve a file, or begin creating a new file. Because it is a relatively slow process, changes are usually not recorded to the hard disk until you explicitly save your work, so program failures (crashes) may obliterate whatever work you have done since the last save. When the program is done it gives control of the machine back to the operating system, which recycles that memory for the next task.

Computer Memory is Lumpy

In digital computers, everything is represented using *bits*, shorthand for "binary digits." Each bit can take one of two states or values. In the early years (1950s to 1970s), when memory was literally magnetic, bits were actually stored as N or S magnetism by pulsing a current through an iron ring. Similarly, punch cards and paper tape either had a hole or no hole in a particular spot. In modern systems bits take the form of high or low voltages sustained in tiny circuits etched by the millions onto chips. Whatever the physical mechanism, each bit is unambiguously one value or the other. There is no "in between" value. Reflecting this, we usually refer to the individual values with dichotomies like "True/False," "on/off," or "1/0."

As the above list of dichotomies suggests, a bit doesn't actually mean anything by itself, but can be used to store one of two values. The meaning arises from the context in which they are used. That is, they must be interpreted, a little like a secret code. This explains why you usually need the same program that created a file in order to change the file. The exceptions arise when the code is widely known, allowing different programs to manipulate the data in the file, as is the case with a plain text file.

Since individual bits store so little info, most of the time we use bits in groups. An 8-bit group is called a *byte*. (Humorous fact: a 4-bit group is called a *nibble*.) The range of meaning possible in a group depends on the number of bits used together. Figure 4.1 uses groups of four circles to illustrate the 16 unique on/off patterns a nibble may have. If you look at it carefully, you'll see that the pattern of filled circles also follows a familiar pattern—starting at the right side, count from 0 to 1; when you run out of digits carry over 1 to the next place (bit) to the left. Repeat. This is counting with binary numbers. It works just like counting with decimal numbers, except binary numbers have a 1s place, a 2s place, a 4s place, and an 8s place, each allowed to hold only a 0 or a 1, rather than the more familiar 1s place, 10s place and so on with digits 0 to 9. Arithmetic doesn't change, just the representation: Look carefully and you'll find that 2 + 5 = 7 here too.

Binary representation of integers

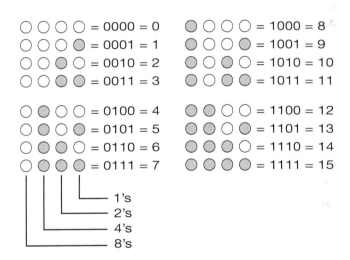

Binary representation of characters

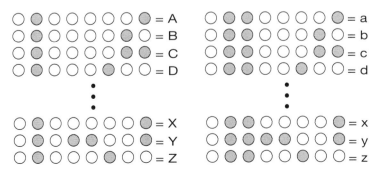

FIGURE 4.1 A few bit patterns and their meanings as integers or text.

Unfortunately, while more bits can store more information, programs are usually built with fixed-size assumptions about the most demanding situation they will encounter—the "biggest number," "number of colors," "maximum number of rooms," etc. These numbers reflect the underlying use of bits to store the information (16 values with 4 bits, 256 with 8, 16 million with 24 bits, etc.). The progression includes $2^{10} = 1024$, which is so close to 1000 that the pseudo-metric notation of a KB (kilobyte) is often used for measuring file size, transfer speeds, etc. Some hardware limits reveal their binary nature through phrases like "32-bit color" displays, "10-bit analogue to digital conversion," etc. If the programmer of your social networking site only allocated 8 bits to store your age, you can't get older than 255! At this time this isn't a problem, but it is worth noting that Microsoft built the original MS-DOS operating system such that the amount of memory a program could talk to directly was only 640 KB and it became a major headache as hardware grew more capable. Today's laptops have about 10,000 times this amount of RAM.

Computer memory is usually accessed, and contents moved or updated, in 8-bit bytes or in multiples of 4 or 8 bytes (32 or 64 bits) at a time. These bigger groupings, called *words*, allow the machine to count to some pretty big numbers, but they're still limited and literal. Humans might accept the notion that the ratio $1/x$ approaches something called *infinity* as x approaches zero, but computers will crash if presented with such information (it's called *overflow*, meaning the number is too big to be represented with the bits assigned to it) because the concept of infinity is an abstraction. It doesn't have a representation in bits.

Computer Memory is Sequential

Unlike human memory, which appears to be associative in nature (your memories of the feel of sand, the shape of beach-balls, and the sound of sea gulls are probably all linked together by connections between your brain's neurons), all varieties of computer storage are organized as sequential storage locations. However, since program instructions and data are all mixed up in memory, it is designed so that any random location can be accessed in the same amount of time, giving rise to the name *random access memory* (RAM). Each word in RAM has a unique numeric address that identifies it. Numbers or text stored sequentially in time (as when typing) are often found sequentially in memory. If we know how to find the first character in a sentence, the odds are the second character is stored in the next byte. This implicit sequentiality is consistent with the way text appears on the printed page or with manipulating a list of values. It's also one of the keys to writing computer viruses (malware) since adjacent areas of memory may be occupied by instructions or data with equal ease and utility.

However, there is no certainty that related information will be stored in adjacent locations. For example, two lines of a CAD drawing that occur in the area we might call the "northwest corner of the kitchen," maybe even touching each other, might be stored anywhere in a multi-megabyte file.

Code: Representing Thought

Computer programming occurs in many forms. In each, it consists of a flexible means of organizing elemental computational actions into sequences or procedures that accomplish information-processing tasks. Because machines take their instructions literally and don't have the opportunity to develop an understanding of your wishes through conversation, this requires careful reasoning about the task and adherence to some language (textual or graphical) of expression during programming. Aside from the (always frustrating) challenges of syntax mastery, programming tools also require us to construct representations, manipulate them in consistent ways, and develop a workable fit between the user (even if it's ourselves) and the information available.

Ivan Sutherland's *Sketchpad* program (1963) allowed a designer to draw and edit simple graphics. In this sense it was a precursor to today's CAD systems, but Sketchpad also allowed the designer to create persistent *relationships* between parts (e.g., end-point matching and equidistant distribution of points around a circle). These features are rarely present in CAD editors today, though modern scripting languages and constraint-based modeling such as that found in some building information modeling (BIM) software begin to reproduce this functionality. When present, the result is an "executable design" that may respond to complex physical, social, aesthetic, or environmental factors and adjust the building footprint, or the façade, or just the ceiling of the lobby, with no added effort.

Though full-featured programing languages can be very complicated, and individual programs can be hundreds of thousands of lines long, the elemental actions used to construct a program are simple and fairly limited. Each program is like a recipe. If you collect the right ingredients (data) and carry out the steps in sequence, you'll get the desired result. The data we select to store and manipulate will constitute the data model for the program; the affordances we create for acting on that data will become the interaction model.

Types of Data

A more detailed discussion of basic data types is provided in the next chapter, but for now what matters is that different pieces of data are converted to patterns of bits using different encoding rules, each creating a different *type* of data, with names like Boolean (for true/false information), integer (for counting numbers), floating point (for decimal numbers), and character or string (for text). Because their encoding rules are different, different types are generally handled separately, though there may be ways of converting between them. Combinations of these basic, or *primitive*, data types can be used to represent more complex data, such as coordinates in space (three floating point numbers), pixels in a photograph (red/green/blue values for each pixel), music samples, etc.

Ultimately, everything gets turned into a number and then a pattern of bits, even text (every character has a numerical equivalent), sounds (volume, by frequency), and colors (intensity of primary colors). This means that complex concepts, especially those involving relationships among parts that cannot be directly represented as numbers, such as beauty or symmetry, are difficult to compute about.

Programmers build more complex structures from these simple elements. For instance, every "album" of music on your computer likely has a title (text), copyright holder (text), year (integer), artist (text), and a list of cuts (each with title (text) and duration (float)). And, of course, there's the music. These more complex structures of data are sometimes called "objects" or "classes"—a nomenclature that spills over into CAD systems via certain programming languages, as well as the Industry Foundation Classes (IFC) used for standardizing architecture, engineering, and construction (AEC) data exchange.

Variables and Constants

Computer programs may have both information that varies (*variables*) and information that remains constant (e.g. the value of *pi*). If you wish to convert someone's weight in pounds to their weight in kilograms, start with the weight (that's a variable), and divide that by 2.20462 pounds/kilogram (that's a constant). It is common to give variables names like *weight_in_pounds* so you can refer to them later. Constants can have names too, or you can use numbers directly in a computation.

Pseudo-code

The following paragraphs will introduce a number of programming concepts. Since most programming is done using text, the examples will be expressed using text, but in order to make them easier to read, they are not written in any actual programming language; they are in pseudo-code.

Assigning Values

In our "pounds to kilos" converter we need to perform a computation to create a new value. Let's call it *weight_in_kilos*. We compute it this way:

   weight_in_kilos = weight_in_pounds / conversion_factor

This is both a statement of truth such as you might encounter in algebra, and a statement of process ("divide the *weight_in_pounds* by the *conversion_factor* and store the results in *weight_in_kilos*"). In programming languages it is the process that matters. It is called an "assignment statement" because it stores or assigns a value to the variable on the left side.

Note that if we had ignored the units in our variable names, we might have written:

   weight = weight / conversion_factor

The statement would be correct in most computer languages. As an algebraic statement of truth it makes no sense unless the *conversion_factor* is 1.0, but as a statement of process it still yields the correct results. You begin with weight (measured in pounds) and you end up with weight (measured in kilograms). They aren't the same number; the weight in pounds is destroyed when the assignment happens.

If this seems a little confusing, you aren't alone. To resolve this symbolic confusion some programming languages use separate symbols to distinguish between assignment and the more traditional "is the same as" or "is equal to" algebraic reading. For example, JavaScript, Java, PHP and C all use "=" for assignment and "==" for "is the same value as."

Parameterized Computations

Computing recipes (algorithms) usually consist of multiple steps. Sequencing is implied by the order in which the steps are listed in the source text file (if it is text) or some sort of dependency graph (in the case of spreadsheets and some visual programming languages). Often, steps involve doing simple arithmetic on variables, but sometimes we need to compute more complex values like the trigonometric *sine of an angle*, a computation that takes several steps itself. Languages usually have a library of such well-defined operations, or *functions*, each of which has a name that can be used along with one or more *parameters* in a statement, like $x = sin(angle)$ where *sin* is the function and *angle* is the parameter. You may be more familiar with function names from cooking, such as dice, chop, whip, simmer, and brown. As with programming recipes, *chop* can be done to onions, carrots, potatoes, etc. The recipe tells us what to chop via the parameter, as in *chop(onions)*.

Functions allow us to condense the text of our program, making it easier to write and keep track of in our heads. In the end we might express a simple soup recipe as

$$soup = simmer(chop(onions) + chop(carrots) + broth)$$

Where sequencing is not obvious, most languages use parentheses to imply "do this first." In this case, you simmer a combination of chopped onions, chopped carrots, and some broth, and call the result soup.

Defining Your Own Functions

Two of the most powerful affordances of a programming language are the ability to define your own unique data and your own functions. Returning to cooking for a moment, we might note that it isn't uncommon to both peel and chop. What if your library doesn't have a function for that? Must we always remember to write down both, in sequence? Couldn't we just say *peel_n_chop(stuff)*? In fact, in much the same way that CAD software allows you to define a "block" (of data) by naming a collection of graphic primitives, we can use a "function declaration" to define a new operation. It takes the form of a vocabulary declaration that defines the list of steps needed to accomplish the action, but it doesn't actually do the task:

```
define peel_n_chop(stuff)
    peeled_stuff = peel(stuff)
    chopped_stuff = chop(peeled_stuff)
    return(chopped_stuff)
end
```

Notice how "stuff" is used. In the first line it is part of the definition, indicating that our *peel_n_chop* function will have a single input parameter, *stuff*. We then use *stuff* anywhere in the function definition where that information should go. Whatever information (or vegetable) we eventually insert for "stuff" when we use our function will be passed sequentially through a peeling and a chopping process, and then returned to the spot where *peel_n_chop* was used. With this definition in place we could write:

```
gespacho = peel_n_chop(tomatoes) + chop(jalapenos) + peel_n_chop(onions)
```

By building up and combining a sophisticated vocabulary of operations, acting on a complex data description, our functions can be made very powerful, but we need a few more tools, chief among them being *conditional execution*.

Conditional Execution

We don't always do every step in a recipe. For instance, a recipe for cooking a roast might say: "If the meat is frozen, allow it to thaw in the refrigerator overnight." Obviously, if it is not frozen, you don't need to do this, but if it is frozen, maybe there is a way to thaw it faster? So, maybe the recipe could say: "If the meat is frozen, allow it to thaw in the refrigerator for a day or thaw for 10 minutes in a microwave on low power." Again, if it is not frozen, you skip over all of this; but if it is frozen, the instructions give you two alternatives but no explicit means of selecting between them except the implied "if you don't have a day to spend thawing the meat." So there are actually two conditions: (1) thawed or frozen; and (2) with time to thaw in the fridge or not. The second only applies if the answer to the first is "frozen." This is called a "nested if" condition. We might write it as

```
If (meat is frozen) then
    if (there is enough time) then
            thaw it in the fridge
    else
            thaw it in the microwave
    endif
endif
```

Conditional execution, or *if–then*, statements generally involve some sort of test (in parentheses in the example above) that produces a *true* or *false* result and some way of indicating the steps that get skipped or executed. As shown in the inner if–then–else statement above, they may indicate both what happens when the test is *true* and what to do when it is *false* (else). The *endif* marks the end of the conditional. Regardless of the conditional test, the next step in the recipe is the one after the corresponding *endif*.

Repetitive Execution

The flip side of conditional execution is repeated execution. What if our language for cooking does not have *chop* in it but it does have *slice*? We could define our own *chop* function as we did above for *peel_n_chop*, by using *slice* repeatedly. One way to do this would be:

> define chop (stuff)
>   slices = slice ¼ inch off the end of stuff
>   slices = slices + slice ¼ inch off the end of stuff
>   slices = slices + slice ¼ inch off the end of stuff
>   slices = slices + slice ¼ inch off the end of stuff
>   slices = slices + slice ¼ inch off the end of stuff
>   return (slices)
> end

This is awkward because it is inefficient to type, hard to update, and worst of all, it is not clear how many slices we need to chop all of *stuff*! We would like the function to work for both carrots and potatoes. In fact, we can express this more succinctly and more powerfully as:

> define chop (stuff)
>   while(length of stuff > ½ inch)
>   slices = slices + slice ¼ inch off the end of stuff
>   endwhile
>   return (slices)
> end

Programming languages usually have multiple ways of saying this (variously referred to using names like *do loop, for loop,* or *while loop*). They are similar to a function in that they define a group of one or more actions that will be repeated. Simple loops just repeat an action some number of times, while others include conditional execution tests made at the beginning or the end to decide whether to do the actions again.

Alternative Programming Environments

The fragments of pseudo-code presented here are meant to convey the linear, sequential character of programming, but aren't written in any specific language. While the underlying nature of program execution remains the same, there are other ways of writing the instructions. The Logo programming environment included a mechanical "turtle." The turtle was part robot, part mouse. Users (usually children) could move the turtle through a pattern of turn-and-advance operations that were captured and then replayed in loops. With a pen attached, the turtle could even draw intricate patterns.

Similarly, many applications allow you to record a series of normal manipulations, such as searches or edits, and then replay them as "macros" or "scripts," a feature which is very similar to the idea of a function described here.

In recent years, visual programming extensions of 3D modeling programs have become popular (e.g., Grasshopper, Generative Components, and Dynamo). In these systems, computational steps are represented by rectangles, with input values on one side and output on the other. Lines connecting the boxes indicate how data flows from one operation to the next, implicitly dictating the sequence of operations. The graphical presentation of such environments has proven very attractive to the visual design community.

Is That It?

Data, combinations of data, sequences of operations, functions, loops, and conditional execution—that's pretty much the basics of constructing a computer program. The obvious gap between this simple description and the sophistication of most tools we use is indicative of complex logic, thousands of man-hours, hundreds-of-thousands of lines of code, and complex system libraries behind such seemingly simple operations as drawing on the screen or receiving mouse input. Still, all that complexity boils down to these fairly simple operations and representations.

Standard Representations

Sequences, functions, loops, and conditionals describe the actions or verbs of computer programs. The nouns are found in the patterns of bits stored in memory. Certain standard representations have emerged over the years and are shared by most computers. Standard representations are essential to the sharing of data between computers produced by different manufacturers and between different programs within a single computer, and they are the fundamental elements out of which more complex representations are built. They make it easier for programmers to move from job to job and for program source code to be ported between differing systems. The representations simply define how to store numbers and text, but there are important subdivisions within those as well.

The existing standards are the product of government and industry experts working together over a period of time. New representations rarely appear as new standards right away. Intellectual property rights and proprietary commercial interests protect them. However, where a single corporate entity dominates a market, such representations often become de facto standards over time as other software developers license or reverse engineer their behavior. Examples include PDF (by Adobe) and DWG (by Autodesk).

Numbers

The most direct representation possible in memory is that of a counting number, where each bit in memory represents a 0 or 1, and a collection of bits represents a multi-digit number. The resulting *binary* numbers have a 1s place, 2s place, 4s place, etc. (each time you add a bit you double the number of unique ways you can configure the collection of bits). An 8-bit byte can store 256 patterns (2 raised to the 8th power), such as the counting numbers in the range of 0 to 255. Two bytes (16 bits) get you from 0 to 65,535.

Negative numbers require a slightly different representation. The positive or negative sign of a number is a duality like true and false, so it is easy to see that if we dedicate one bit to remembering the sign of the number we can use the rest of the bits for normal counting. While preserving the number of unique representations, they are divided between the positive and negative portions of the number line. One such scheme in common use is *two's complement* encoding under which an 8-bit byte can store a "signed" value between −128 and +127 or an "unsigned" value (always positive) between 0 and 255.

The positive and negative counting numbers are called *integers*. Such numbers have exact representations in computer memory. The range of values they can have depends on the amount of memory allocated to store each one.

A chicken counting her eggs could use an integer, but an astronomer figuring the distance to the nearest star needs a different representation because it takes too many digits to record astronomical distances even if they are measured in kilometers rather than inches. That's why *scientific* or *exponential* notation was invented! Using this scheme the distance to Alpha Centauri can be written as 4.367 light years, and since a light year is 9.4605284×10^{15} meters, that makes a total of 4.13×10^{16} or 413×10^{14} meters after rounding off a little. Interestingly, that last way of writing it uses two much smaller integers (413 and 14), at the expense of rounding off or approximating the value. Using negative exponents we can also represent very small numbers. Because these kinds of numbers have a decimal point in them, they're generally called *floating point* or *real* numbers. Some of the bits store the exponent and some store the significant digits of the numeric part (called the *significand*). In a 32-bit single-precision floating point number, 24 of the bits will be used for the significand, and 8 bits will be used for the exponent. That translates into around seven significant digits in a number ranging in value from -10^{38} to 10^{38}. To do math with real numbers is a bit more complicated than doing math with integers, but dedicated hardware takes care of the details. Note that while they can represent very large or very small values, floating point numbers are necessarily approximate.

There is a difference between a number (e.g., 12) and a measurement (12 meters) that includes a unit of measure, because the standards for computer representation only cover the number. The number can represent a measurement in any unit of measure. The standards guarantee that numbers shared between programs represent the same values, but there is nothing in the standard representations that says what units the values are measured in. Unfortunately, the units of measure are often implicit in the software (CAD programs in the United States default to inches or sometimes feet), or left up to the user (as in a spreadsheet). In 1999 the simple failure to convert a measurement from imperial to metric (SI) units when sharing data within an international design team led to the loss of the multi-million dollar Mars Climate satellite (Hotz 1999).

Because units are usually implicit in CAD software, and exchange formats generally transfer numeric coordinates rather than unit-oriented dimensions, architects sometimes find that drawings exported by a consultant need to be scaled up or down by 12 (to convert between feet and inches), or 1000 (to convert between meters and millimeters), or by 2.54 (to get from inches to centimeters or back).

The Fuzzy Edges of Floating Point Coordinate Systems

The important point about the range of a representation is that it cannot capture the full splendor of reality. If you establish a coordinate system for our solar system (centered on the sun) using single-precision real-number coordinates measured in inches, coordinates are precise to about eight digits. For points near the sun, that's accurate to a tiny fraction of an inch, but for positions 93 million miles (5.9×10^{12} inches) away from the origin, say on the surface of the Earth, that means points 10^4 inches (833 feet, 40 meters) apart have indistinguishable coordinates. Not a big problem for architects who usually pick an origin closer to their projects, but a challenge for space scientists and molecular engineers. Larger memory allocations (64 bits per number rather than 32) reduce the problem, but it never completely goes away.

Which is Best? Integers or Floating Point Numbers?

Integers are generally faster to compute with, but not useful for things like spatial coordinates, because we often need fractional coordinates, and we may need big coordinates.

The curious thing in this situation is that floating point numbers, which allow us the most precise-appearing representation (e.g., "3.1415926" for *pi*, or "1.0") are using an internal representation that is an *approximation* to the value, while the numbers we associate with approximation (e.g., "37 feet" or "12 meters") are actually exact representations (if you rounded off the measurement when you wrote it down, that approximation is on you—the internal representation of the number is exact).

By convention, integers generally appear without a decimal point (so "37.0" is a floating point number, while "37" is an integer), but this is not a requirement.

Text

Humans write text in many ways—left-to-right, right-to-left, top-to-bottom—and we use many different symbol systems, some with a handful of unique symbols, like American English, and some with thousands, like Japanese Kanji. This complicates the creation and entry of text in ways that Western Europeans and their American cousins don't always appreciate, but since American companies dominated computing in the 1980s we started out doing it their way. To simplify things and make it manageable, the early systems ignored all text except left-to-right, top-to-bottom. This reduced the complexity down to just two problems, analogous to typing: picking the individual characters and figuring out when to start a new line of output.

ASCII

Remember, bytes store patterns of on/off, so they don't actually store a text character. The link between the pattern and the meaning is one we create. It is arbitrary, but it can be rational. To simplify the problem, early developers ignored much punctuation and many letters. Given a list of acceptable characters, they created mappings from on/off patterns of bits in memory to the selected characters—very much like setting up a secret code. A few of these, from the *American Standard Code for Information Interchange* (ASCII, pronounced "ass-key"), are shown in Figure 4.1, where you might note that "A" and "a" are not the same character, and that interpreting the bits of "B" as an integer yields a larger number than "A"—a fact that helps with alphabetizing.

The *appearance* of the letters when printed depended entirely on the printer—there was no such thing as a *font*. Even worse, different companies (DEC, IBM, CDC) came up with different mappings, so an "!" (in EBCDIC) on an IBM mainframe had the same bit pattern as a "Z" (in ASCII) on a DEC mini-computer. This significantly complicated the process of copying files from one computer to another, but since much data stayed in one place it wasn't a huge problem. When the PC revolution began and people wanted data on their PC that started out on their company mainframe computer, the incompatibilities between systems became both noticeable and painful.

The rise of the PC saw the demise of all the encoding schemes except ASCII. ASCII rigidly defines 128 characters, including both upper-case and lower-case letters, numerals, and a variety of punctuation, plus some characters like "form-feed" (to start a new page), "line-feed" (move down a line), and "carriage return" (move back to the start of the line).

ASCII uses an 8-bit byte to store each character, so there are 256 possible characters, but the standard only defined the first half of them. The eighth bit was originally set aside to provide a way to validate character transmission between systems, but robust hardware rendered that use unnecessary and vendors began to use the "upper 128" characters for other things. Different vendors (IBM, Microsoft, Apple, etc.) filled in the missing characters but not in the same way, so while ASCII is a standard, it is a flawed standard in practice. The flaws appear most obviously today when you use certain characters in email or blog posts, characters such as curly (or smart) quotes, accented characters (éøü), em-dash (—), etc. The vacuum of undefined space had some positive effects too; in the 1980s those upper 128 characters allowed researchers at the University of Hawai'i to develop a script system for recording Pacific Northwest Native American languages and stories, helping preserve a language that was rapidly disappearing (Hsu 1985). But ASCII is no good if you want to use a writing system from the Middle East or the Far East. Those users literally could not write things down in a familiar and standardized script on a computer until unicode came along.

Unicode

Unicode was created to address the shortcomings of ASCII and its various extensions. The goal was to provide a unique character encoding for every writing system in the world (Unicode 2016). Because memory and storage space have been scarce resources, and in order to make the new standard backwards compatible with ASCII, unicode actually comes in three slightly different variations: using 1 byte (utf-8, very similar to ASCII), 2 bytes (utf-16), or 4 bytes per character (utf-32). Individual characters, called *code points*, are assigned in language-oriented groups.

End-of-Line

While unicode resolves the character representation problem, there is another problem with text files. The ASCII standard doesn't specify how the end of a line is marked in a text file. With the development of word processors and automated word-wrap, this has morphed into a question about how the end of a paragraph is marked too. Unfortunately, three incompatible answers evolved in the industry:

- Beginning with the observation that when you type on a typewriter you push the carriage-return lever or press the Return key, creating a "carriage return" (CR) character to start a new line, some systems use a CR to mark end-of-line.
- Again observing the typewriter or printer, but noting that the start of the new line is both at the left edge and down one line on the page, some systems mark

an end-of-line with a combination of a CR and a line-feed (LF) character, sometimes written CRLF.
- Printing bold or underlined text used to require printing on the same line twice, which required doing a carriage-return without a line-feed. Such systems opted to mark the end-of-line with a single LF, which caused both physical transformations.

Microsoft Windows uses CRLF. Unix (and thus OS-X) uses LF. The original Apple Macintosh used CR. If you have ever opened a text file to find it unaccountably double-spaced, or with some odd invisible character at the start of every line after the first one, or where each paragraph is one very long line or the entire file turns into one giant block paragraph, you've encountered the side-effects of this problem. HTML rather brilliantly dodges the whole mess by lumping all three into the general category of "white space" and treating them the same as a space character.

Characters, Glyphs, and Fonts

The name of each letter (e.g., "upper-case A") is distinct from the graphic mark (or *glyph*) with which it is printed. A collection of glyphs is called a *font*. In the early days text was printed by striking a character image against an inked ribbon, making what we now call "plain text"—all characters were the same width (ten characters per inch); 80 characters filled a line; and there were six lines per inch vertically. All upper-case "A" characters looked the same, but their font might vary from terminal to terminal or printer to printer, depending on the manufacturer, not the computer or file. In the 1980s the appearance of raster output devices such as dot-matrix and laser printers changed all that, giving us proportionally spaced text of many sizes and appearances that flows into variably spaced lines and paragraphs.

Even today, font data largely remains separate from the character data, and because fonts can be protected by copyright, not all fonts are available on all computers, which means exchanging files (both text documents and web pages) between operating systems can cause unexpected font substitutions. When an alternative font is used, lines of text usually turn out to have different lengths. This causes the edges of fully justified paragraphs to not line up; shifts text on slides and in notes or dimensions in drawings and illustrations, possibly causing overlaps with other line-work or graphics, etc. Even though fonts may now be attached to web pages and can be embedded in PDF files, intellectual property restrictions on some copyrighted fonts remain a challenge.

Files and Directories

The different virtual devices that your computer can become are encoded in the different programs it has stored on the secondary storage system—the hard drive—of your computer. Similarly, the different documents you might compose with your word processor are stored separately so you don't have to complete one document before starting another. Instead, you *save* and *open* them as needed. Each document or application is assigned space in secondary storage called a *file*, and files are organized into *directories* or *folders*.

As a *container*, each file is standardized and cataloged by the operating system, but the *contents* of the file, the pattern of text and numbers called the file *format*, is specific to the program that created it, and those programs probably come from different companies. Thus, your word processor can't make sense of your CAD data and vice versa. File *extensions*, suffixes to the file name such as .doc or .html, are commonly used to indicate which program goes with which file.

The one universal file format is plain text (usually ASCII). Most programs can read and write text files, though saving data to plain text may cause important features to be discarded. Chapter 8 explores the question of representation in more depth.

Summary

Patterns of bits stored in files represent text and numbers, as well as program instructions. Programming languages, with their data types, variables, branches, functions, and loops, give us the means by which to express and organize sequences of operations to do useful work, creating entire new functions and programs, or scripts to run inside existing programs. Standardized representations for text (ASCII and unicode) and numbers (integers and floating point numbers) provide portability, but only cover fairly simple data types. History and commercial interests have made more sophisticated data exchange more difficult, but network connectivity is encouraging the development of exchange standards. Chapter 5 will explore standards emerging in the AEC industry.

Suggested Reading

Sutherland, Ivan. 1963. SketchPad: A man–machine graphical communication system. *AFIPS Conference Proceedings 23*: 323–328.

References

Hotz, Robert Lee. 1999. Mars probe lost due to simple math error. *Los Angeles Times*, October 1, 1999.
Hsu, Robert. 1985. *Lexware manual*. Manoa: University of Hawai'i Department of Linguistics.
Sutherland, Ivan. 1963. SketchPad: A man–machine graphical communication system. *AFIPS Conference Proceedings 23*: 323–328.
Unicode. 2016. What is unicode. www.unicode.org/standard/WhatIsUnicode.html

5
DOING WHAT DESIGNERS DO

In the 1950s and 1960s the apparent similarities between the adaptable abstract computing machine and the human mind led many to expect the rapid emergence of human-quality intelligence from computers. They were disappointed. Human brains and human thinking seem to be much more complicated than we thought, and the particular strengths of machine intelligence are perhaps of a different sort. While asking how designers *think* has not been immediately productive, asking what designers *do* has been. Focusing on observable information-processing behaviors—the creation, sharing, and modification of calculations, drawings, and other documents—has produced the current state-of-the-art in design practice.

In the shift from centralized mainframe systems to desktop productivity tools that accompanied the microcomputer revolution, interactive graphics became an important feature. Starting with the commercial success of the Apple Macintosh, the WIMP (windows icons menus and pointers) interface became the standard for both operating systems and programs. The availability of inexpensive graphic displays and input allowed drafting and rendering applications to develop. In the last two decades computers have advanced beyond the role of personal engineering and drafting platform to that of 2D and 3D visual arts platform, dramatically improving their utility and appeal across design fields. Network connectivity has made collaboration and data sharing both routine and efficient, while providing access to information worldwide. The combination has driven computing to the center of most design practices, though analysis and documentation remain an important part of workflow.

To understand the state-of-the-art, we must consider architectural applications in sketching, drawing production, building information modeling, automation, analysis, fabrication, and even operation—all aspects of buildings where design computing has had an impact. First, however, due to the centrality of geometry and form in modern design software, we need to review 2D and 3D software principles.

Graphics

Architects aren't the only designers who draw. Software vendors, seeking to spread the cost of developing the hundreds of thousands of lines of code and appeal to as many users as possible, have made their products as generic as possible. This has meant, until fairly recently, that architectural design tools could be shared with or repurposed from fields as diverse as movie animation, engineering, aircraft design, and jewelry design. In particular, the high-quality rendering needs of Hollywood animations and special effects have combined with the purchasing power and visual thirst of video-game players to revolutionize computer graphics. Sophisticated software and data structures for manipulating and rendering 3D objects have been developed, widely shared across the computer graphics industry, and turned into special-purpose graphics hardware which is now found in almost every modern desktop, laptop, tablet or smartphone, as well as dedicated game consoles.

There is more to a building than its geometry, more involved in creating, servicing, and using it than how it looks. The digital representation of a building is more complicated than shape, but as graphics has long been a key part of design it does provide a starting point. There are currently two broad strategies for recording, storing, and manipulating 2D computer graphics (vector and raster) and four for 3D (wireframe, surface, solid, and voxel). We'll look at those in this chapter, returning in Chapter 8 to consideration of their architectural application.

The international standards for representing numbers and text are precise to the level of the bit. More complex data, including 2D and 3D graphics, often follows very standard conceptual conventions, but these do not extend to the level of precision we can find for elemental data types. As the emergence of ASCII as the dominant representation for text was aided by the market dominance of the personal computer, several other de facto standards have emerged in the area of graphics. These are built on top of simpler standards, becoming increasingly complex, and increasingly difficult to negotiate broadly, as they rise.

2D Graphics: The Difference Between Vector and Raster

Imagine communicating a drawing to a colleague over the phone. Speaking "architect to architect," you might say: "There is a brick-veneer stud wall on the north and east, with an 18-inch (500 mm) double-hung window centered 36 inches (1 m) away from the north wall. The 2-inch (50 mm) walnut inlay around the perimeter of the oak strip floor turns the corner inline with the southern edge of the window." The communication is precise and clear, but contains lots of architectural jargon and lots of information about intent. This is a "high-level" architectural representation. It relies on knowledgeable architectural interpretation. This is the level of interaction that BIM programs aspire to.

Now, imagine communicating a drawing to a less knowledgeable worker over the phone. Speaking "architect to drafter," you might say: "Using a double-line spacing of one-quarter-inch (10 mm), there is heavy double line 1 inch (25 mm) from the left edge of the paper. It meets another heavy double line 1 inch (25 mm) from the top. One and a half inches (40 mm) from the top is a short heavy horizontal line to the left. Two and a quarter inches (60 mm) from the top is another short heavy horizontal line to the left. Using a double-line spacing of 1/16 inch (2 mm), two and a quarter inches (60 mm) from the top edge, a thin horizontal line is drawn. It meets another light line drawn two and a quarter inches (60 mm) from the left edge." (Such descriptions become quite long—this one has been abbreviated.) The communication is all about the graphics, free of architectural jargon. This is an *object* or *vector* graphics representation. It relies only on a precisely defined, if limited, drawing vocabulary and a coordinate system. Drawing or CAD programs are vector-based.

Finally, imagine communicating the drawing to a colleague over the phone, but rather than describing each of the graphical or architectural elements out of which the drawing is constructed you might just slip it into your fax machine and press *send*. The fax machine doesn't know anything about architecture or drawings. It just scans the page using a grid of sample points. Each sample is either white or black. The fax machine scans the page left-to-right and top-to-bottom, generating a series of tones that communicate the sample values it detects. The second fax machine, miles away, prints dots on the page whenever it hears the black tone, and in almost no time at all ... out pops a finished drawing. No disciplinary jargon is used. Drawings and text can both be sent. No explicitly architectural or graphical information is included, but the drawing is kind of rough or *pixelated* in appearance. Figure 5.1 illustrates this dichotomy for a simple circular shape.

The "fax" scenario describes *paint* or *raster graphics*; the name derives from the pattern of beam scanning in old cathode-ray-tube (CRT) televisions. The image consists of a grid of individual *picture elements* or *pixels*. The quality of the resulting image depends on the number of pixels in the image (*resolution*), as well as the number of colors each pixel can have (*color depth* or *color resolution*). Confusingly, the term "resolution" is often used to reference either pixels per inch (ppi) or the total number of pixels horizontally or vertically, leading to such odd sentences as: "A 3 inch by 5 inch (7 cm by 13 cm) image with a resolution of 150 ppi will have a resolution of 450 by 750 pixels." Paint programs use raster graphics.

To store or display a raster image you must know the individual color values for each pixel of the image. That means a blank image has as much data in it as a complex one, and files may need to be quite large in order to capture detail. Reducing spatial resolution and reducing the amount of color information per pixel keeps the file size down. Mathematical treatment of the data may allow you to compress the file, saving space in storage and time in transmission. When you want it back you simply decompress it. Compression/decompression schemes are called *codecs*. If decompression gives back exactly what you started with it is a *lossless* codec (TIFF, PNG, etc.). If compression involves simplifying the file or removing detail that is not restored during decompression, it is a *lossy* codec (GIF, JPG).

For sketching and thinking it is important that the drawing tools do not intrude on the designer's thinking. This is one of the features of pencil and paper that remains compelling. A paint program's individually editable pixels, especially when combined with a touch-sensitive screen or tablet, go a long way towards recreating that simplicity. While the tactile drag of a soft pencil point over rag paper may be missing, software can visually simulate the behavior of a watercolor wash or the irregular marks produced on toothed paper.

Using a raster approach does have drawbacks beyond the feel of the pencil on the paper: You can only zoom in so much before the pixel nature of the image becomes obvious; editing is largely restricted to copying groups of pixels from one place to another; changing the line-weight or color of an existing line is painful; and the data does not lend itself to analysis tasks that need to know lengths, distances, or names. However, since the raster grid gives a color to every pixel and accounts for all parts of the image, color-coding can make computation of some information, e.g., rentable area on a shared office floor or the shaded area of a plaza, as easy as counting pixels of a particular color.

2D Architectural Drawing

Pencil and paper can be used both for free-form sketching and careful drafting. When interactive drawing programs were developed, those using a vector- or object-based representation were generally more supportive of drafting tasks.

Returning to the hypothetical phone conversation with the drafter, if you ask them what they are doing with their tools you probably won't be told that they're recording ideas. You might be told that the lines represent cuts through walls or the edges of materials, but most likely you'll be told that they are "drawing up" the design, working out the geometrical implications of design decisions in a precise way. Such drawings are constructed as miniatures, carefully scaled and abstracted representations of the real world. In the 1980s the American Institute of Architects (AIA) standard contract allocated 40 percent of fees to this task, which involves care, adherence to conventions, and frequent modifications, so it should not be a big surprise that the principals of firms were interested in efficient digital ways of carrying out this activity. This is the market that 2D drafting addresses—creating, editing, and managing carefully constructed line drawings.

Adopting and abstracting the drafter's view of drawing (and the traditional tools of straight-edge, triangle, compass, and French-curve), CAD systems treat a drawing as a collection of marks of different sorts, including straight lines, circles, text, and so on. The different kinds of marks are called *primitives*, and each instance of a primitive is distinct, controlled by critical points stored as Cartesian (x,y) *coordinates*. As an improvement over manual drafting, these coordinates are usually created using real-world dimensions, which eliminates one source of errors during their creation (scaling is done during *plotting*).

If the user has a mouse or graphics tablet for input, it may be desirable to correct their imprecise human input through the software, ensuring that lines are exactly straight and perhaps horizontal or vertical. Sometimes it will be important that a new line meets up exactly with the end (or middle, etc.) of an existing line. Using the mouse's coordinates it is possible to search the existing list of instances to find the nearest endpoint. Those coordinates can then be used as the endpoint of the new line, ensuring that they touch precisely and completing an *endpoint snap*, just one of many object snaps available in drafting software. Notice that the *snap* does not create a persistent relationship between the new endpoint and the existing line.

A drawing can thus be stored and edited as a *list* of lines, circles, arcs, and text, using real-world coordinates. *Graphic attributes*, such as *color*, *line type* (dashed, dotted, etc.) and *line-weight* (thin, wide), are associated with these lines.

There is also a *layer* associated with each drawing element. Layers are an important data-organizing attribute. The drafter can record, via the layer name, what a given line represents (interior wall, edge of roof above, steel structure, to be demolished, etc.), or what the material is (brick, concrete, etc.). By turning layers on and off, different plan drawings (existing, new, reflected-ceiling, framing, etc.) could be produced without duplication of drawing effort. Of course, significant energy went into defining and conventionalizing layer names so that data (and workers) could be exchanged between projects and consultants (NIBS 2014). To this day, large organizations such as government agencies and large manufacturers require work to be done using their particular layer-naming scheme.

A notable characteristic of many construction drawings is that identical groupings of primitives often appear in multiple places, windows in elevations, doors in plan, etc. CAD software allows the user to create collections of lines, often called *blocks* or *symbols*, give them names, and add or insert them repeatedly as needed to speed up drawing work. When the master copy of a symbol is edited, all instances are changed as well.

While the CAD data only indirectly represents a building, the existence of underlying coordinates means we can compute the real-world length of lines or areas of closed shapes. Using appropriate layer names we might indicate materiality (brick, wall board, etc.), making it possible to do material takeoffs. Symbols, which often represent manufactured items such as windows, sinks, or toilets, can be sorted by name and counted for furniture, door, and window schedules, in contrast to copy-and-paste line work. In short, and in contrast with raster representations, a vector drawing begins to represent the building in a *computable* fashion.

Storing Drawing Data

Vector graphics started out focused on the kinds of graphics that can be constructed by moving a pen around on a piece of paper, much as a draftsman would. They were called "vector graphics" because they were made of line segments or simple geometric primitives such as circles or arcs. Storing such drawings is simply done by making a list of the individual primitives, their attributes, and coordinates. The more complex the graphic, the longer the list is and the bigger the file. Primitives are added to the list as the user draws them. Symbol definitions might be included in the file or kept in a separate library. Attributes such as layer name and line weight can be interpreted during plotting as a pen change, or a line can be stroked

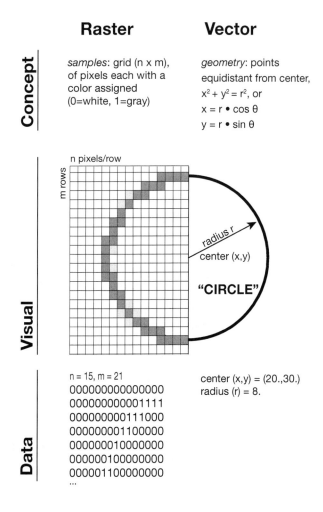

FIGURE 5.1 Vector and raster representations of a circle, in visual, conceptual and data form.

twice with a little offset to make it fatter. Objects on layers that are marked as "off" can be skipped.

There were efforts to standardize this kind of representation. The SIGGRAPH and GKS metafile projects sought to standardize graphic imagery (Henderson and Mumford 1990), but these efforts were largely eclipsed by the overwhelming successes of PostScript for printing and the Autodesk Drawing Exchange Format (DXF) for architectural drawing data (Fallon 1998).

Because the early CAD systems were sold to multiple disciplines, each with a distinct vocabulary, they provided simple geometric primitives in order to serve a broad market. Building "walls," if available at all, might be created by discipline-specific add-ons, extensions of the basic software that drew walls as two parallel lines with mitered corners and a T at junctions. While such extensions facilitated drafting, they were usually incomplete in some way—when you added a window you had to make a hole in the wall first, and when you moved the wall you had to remember to move the window too. That's because they were not really walls or windows—they were just lines. The understanding of them as architectural objects happens outside the software, in our heads. Only in the last decade has software become widely used that truly understands walls in an architectural way.

Vector graphics can be enlarged or reduced without loss of information, but raster graphics become visually boxy if enlarged enough. If a raster graphic is scaled down to a lower resolution, detail is permanently lost and cannot be restored. Almost all modern displays and printers are raster in nature because of the speed and flexibility that this technology makes available. While architects use drawing in multiple ways, the computer graphics industry distinction between raster and vector graphics has become associated with the difference between sketching and drafting. Paint programs offer the unconstrained fluidity of sketching, while vector-based drafting software offers the precision and detail required for construction documents in a compact file format.

3D Graphics: Data + Algorithm

Two-dimensional vector, or "object," graphics are usually employed to capture the precise kinds of data that are needed to create construction documents for buildings, landscapes, etc. This is why drafting systems have such complicated snaps and grids, and even sometimes require the numeric entry of data.

The world of three dimensions, on the other hand, is more often used during the early design process to assess and present the visual character of different design ideas and to answer questions such as "Can we see the garbage dumpster from the deck?"

Oddly, 3D representations are usually simpler than their 2D cousins, at least in terms of the number of primitives they support. They may not include text primitives and rarely include dimensioning tools. Instead, they focus on making it easy to create and view a 3D representation of the design via renderings of simulated views.

While they may use simpler data, three-dimensional systems are very different from their two-dimensional counterparts in that the 2D image you see on the screen changes dramatically as you manipulate the viewpoint, adding substantial complexity to use of the program. Where the display list data in a 2D system directly represents the drawing, the model data of a 3D system is only the raw material from which the screen image is made. In fact, you never see the 3D data, only 2D representations of it. The data is delivered to the screen through what is called a *graphics pipeline*, algorithms, and (often) hardware that convert 3D representations into 2D imagery.

3D Representations

Prompted by academic demonstrations in the early 1960s (Sutherland 1963), a few large architecture, engineering, and construction (AEC) companies developed their own 3D software (Fallon 1998), as did a number of universities, so that 3D software began to be generally available on mainframes in the 1970s. These programs were largely controlled by commands entered at a keyboard. Monochrome line drawings were the main product, either plotted in batch mode or displayed as static images on simple graphic terminals, requiring many seconds or minutes per frame to complete.

Two technological advances disrupted this situation in the 1980s. The first was the appearance of standalone color raster displays, enabling the display of multicolor shaded images as part of mainframe computing. The second was the introduction of microcomputers. While text-only at first, or limited to very crude graphics, it wasn't long before color raster displays and microcomputers merged and the mainframes faded into memory.

As with 2D CAD, most 3D systems are editors and most models are simple lists of primitives, but with two differences: First, the implementation of general-purpose hidden-line and hidden-surface removal means that it is possible to produce traditional 2D plan, section, and elevation views from 3D models; and second, advances in 3D information modeling mean that models can begin to maintain the semantic content that designers recognize ("walls," "windows," etc.). Unfortunately, most hidden-surface processes produce raster images, limiting their utility for automated vector drawing extraction, and the remaining techniques don't often render line-weights with the finesse expected in traditional practice. On the plus-side, color renderings and animations are now possible and 3D models can be used as the foundation for energy and lighting analysis, laying the foundation for the more sophisticated building data management systems (BIM).

There are four main categories of 3D data, illustrated in Figure 5.2. The first three can be thought of as extensions of vector graphics occupying the geometric domains of 1D lines, 2D surfaces, and 3D solids. The fourth fills the 3D niche corresponding to 2D raster data: volumetric elements, or *voxels*, in a 3D grid. A fifth form, point clouds, consisting of (sometimes colored) arbitrary points in 3D space is also emerging, driven by laser scanning technology.

Wireframe

If the data consists simply of lines in space it is called wireframe (a term which may also be used to refer to a "see-through" *drawing* and occasionally to any drawing done only with lines). Wireframe data is quite easy to create and maintain and can be converted easily from 2D data, such as that of a 2D drafting system, by just adding an extra coordinate or extruding lines from 2D into 3D. However, as the name implies, it is not possible to view the model as if it were solid since opaque surface data simply isn't present. Nor is it possible to prepare shaded views of wireframe data. Once common, software limited to wireframe is now rare, though

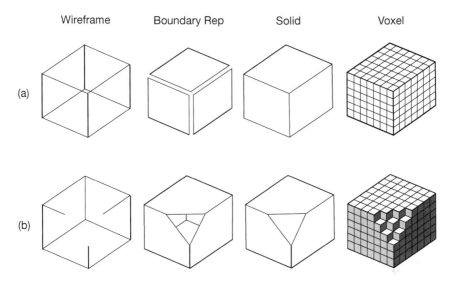

FIGURE 5.2 Standard 3D representations: (a) Wireframe, Boundary Representation, Solid, and Voxel, (b) after removing the near corner.

modern 3D software often uses wireframe rendering (rendering *as if* the data were wireframe) during editing because it is computationally fast. Most programs use surface (boundary representation) or solid data models.

Boundary Representation

Data consisting of two-dimensional surface elements or polygons, defined in a 3D coordinate system, is often called a surface, boundary representation (*b-rep*), polygon mesh, or polyhedral model. This is because, like a cardboard or paper model of a shape, it achieves its form by describing the surface faces or facets of the 3D shape. Because b-rep models contain explicit surface information, they may be used to produce renderings: hidden-line images and surface-shaded images. More complex renderings displaying surface reflectivity and texture can also be computed from b-rep models. For architects and other form-oriented designers, this is often as complex a system as is needed. The problem with this kind of data is that their geometries need not be real. For instance, it is perfectly legal to create a single polygon floating in space. Since polygons are (by definition) one-sided and have no thickness, they have no volume! It might look like a piece of paper, but a stack of them does not fill up a box! They may work for rendering, but you couldn't manufacture (or 3D print) such a shape because it can't exist in the real world.

A b-rep model can represent a very real-looking scene, but there are challenges. If you've visualized a b-rep model as a cardboard box, you are on target. Now imagine representing a hillside, face, car fender, or a teapot with that same cardboard. These irregular shapes are common in the world, but they do not reduce to abstract shapes such as spheres or cylinders any better than they reduce to planes. Two strategies have been developed to work with such shapes: rendering better and modeling better.

The rendering strategy works like this. Approximate irregular forms with flat polygons. This *tessellation* will cause them to be rendered with slightly different shades in the image, but since rendering is raster we can then work to smooth out the resulting color transition along the edge(s) joining adjacent faces. This approach, generally called *shading* or *smoothing*, has been around since the 1980s and is used extensively in the computer graphics industry; it is supported in graphics hardware on most PCs. These algorithms allow us to render even crudely faceted models so they look smooth most of the time, and are common features in video games and immersive online environments. They make the model *look* better in renderings, but if you 3D print it or mill it with a CNC router, the faceted nature or the geometry data will be apparent.

The modeling approach replaces the flat polygon faces with something better: non-uniform rational B-spline (NURBS) surfaces (Figure 5.3; Foley *et al.* 1990). In a NURB surface each polygon is treated as a *patch* or fragment of a larger surface. Each patch is divided into a grid of orthogonal curves, each of which is a composite of several polynomials. The infinity of the patch is tamed by a standardized mathematical representation and some unique parameters. The mathematics behind the representation enable a fairly small set of control points and related values to be used to compute the coordinates of any point on the surface to an arbitrary level of precision. This formulation assures that (a) edges

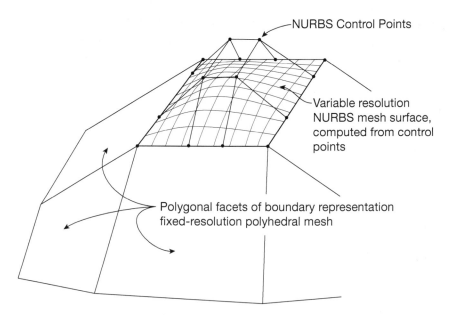

FIGURE 5.3 Fixed-resolution B-Rep surface compared to variable-resolution NURB surface.

align (no slits or gaps), (b) slopes match (no creases), and (c) curvature matches (human eyes are incredibly sensitive to light changes!). An appropriately low-resolution *polygon mesh* can be generated from the mathematical surface as needed and used with the smoothing algorithms mentioned above to deliver the right visual experience. In addition, the geometry can be used to 3D print or cut a shape that looks and feels correct.

Solid Models

There is no difference in the appearance of b-rep and solid models, so either can be used for rendering. Most of the difference lies in what you can do with the model. When the basic shapes that the system works with are 3D and the software is able to perform operations such as Booleans (e.g., union, difference, and split) on those shapes, the system is then said to be a *solid modeling system*. Surface or b-rep models aren't quite enough. For example, six squares, glued together along their edges, make a cube-shaped box. It might look "solid" but if you run it through a band saw you'll see the cavity inside. Run a cube of timber through the band saw and you have two solid parts, and no cavity. The cutting operation corresponds to the Boolean operation. To complete it, the system must know quite precisely what is inside the boundary shape and what is outside. Curiously, much the same information is needed to 3D print a shape.

The basic boundary representation can be extended to define solids. A b-rep that satisfies a relationship called Euler's Formula is considered solid—also *closed* or *watertight*. Boolean operations, defined in terms of a set of elemental Euler operators, can be used to compute unions, differences, and intersections between two such objects, while assuring that the geometry remains that of a solid. Using NURBS as the surface elements, the geometry of projects from Peter Eisenman's Holocaust Memorial in Berlin to Frank Gehry's Guggenheim Museum in Bilbao, can be modeled, rendered, and potentially fabricated (at least in model form) reliably.

Solid modeling systems are commonly used in industries that actually manufacture objects from CAD data, such as the aerospace, automotive, and consumer-products industries, because the software can guarantee that the shapes it produces can exist in the real world.

3D CSG Models

One type of solid modeler, called a constructive solid geometry (CSG) modeler, goes a step further. Using geometric primitives that represent both the raw material (e.g., a block of aluminum) and the "negative" volumes of cutting, drilling, or milling operations, a CSG modeler tracks the history of edits applied to the initial form. This information can be used to completely define the part geometry, as well as the manufacturing steps needed to make it. In addition, the part's mass properties (volume, weight, centroid, etc.) can be computed, and it can be virtually combined with other parts into functional assemblies such as an automobile engine whose operation can be simulated to test dynamic clearances, etc. As the description may suggest, it is not commonly used in architectural design.

Volumetric Models

The 3D equivalent of raster data is *voxel* data or *point clouds*. Until inexpensive laser-scanning systems began to appear in the AEC workflow, this data was largely restricted to diagnostic medical imaging. Think of them as (possibly colored) points in space. Voxel data will follow a grid pattern while point clouds simply provide points of color in space. Though there are no surfaces, with enough points you can render the scene from different points of view, and if you work at it you can identify and extract elements of geometry (pipes, surfaces, etc.).

While not common, such data is increasingly used in AEC workflows to establish *as built* conditions in projects such as refineries or buildings with complex mechanical systems, though it is difficult to reconstruct surfaces (or building elements) from the point data.

The Difference Between Geometry and Buildings

There are ongoing efforts to create robust and flexible standards for representing building models in academia as well as national and international standards-setting organizations, though none appear strong enough to be widely adopted yet. We'll talk about these efforts further in Chapter 8, but for now let's just get a sense of the problem by considering a couple of related issues: topology and geometry.

The visual aspects of a building, both interior and exterior, can be captured in a 3D surface model. Such a model may include visual textures for rendering, but is unlikely to have information about the material, structural or thermal properties of the assemblies it helps visualize or even their geometric details. This model is all about visual geometry and texture.

Architecture, on the other hand, is about space, not the objects in the space or the walls around it. What goes on in the space matters, so one common tool for attacking a design problem is a "bubble diagram," as illustrated in Figure 5.4a. It identifies spaces and their relationships, but not their sizes or (necessarily) walls. The list of spaces and their connections establishes what mathematicians call a *planar graph*. Between the *nodes* (spaces), the connecting *edges*, representing connections, imply doors, and the doors imply walls. After squaring up the rooms and providing preliminary dimensions, the *dual* of the graph emerges as a preliminary floor plan in Figure 5.4b. This is *non-manifold* geometry. Like a literal soap-bubble cluster, it uses surfaces to divide the world into volumes or spaces. Moving a surface affects both adjacent volumes, so the representation is inherently spatial and topological. However, viewed as walls, the surfaces face both ways and often form T-junctions, two conditions that traditional solid models do not allow because it confuses the meaning of "inside" and "outside."

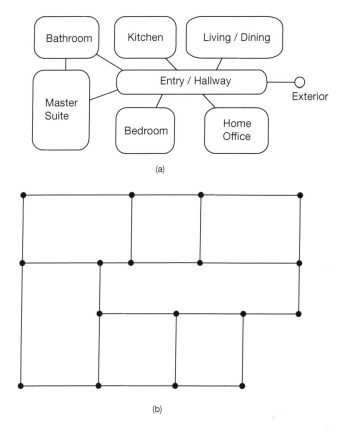

FIGURE 5.4 Architectural bubble diagram and one possible floor plan.

Conversion Between Representations

The similarities and differences between 3D wireframe, boundary-representation, and solid models can be used to illustrate a general problem related to conversion between representations: There is often directionality to such conversions, with one direction being easy and the other hard. As shown in Figure 5.5, given only the 12 *edges* of a cube, there is no way to know if they represent a cube, or a box, a tube, or a wire sculpture. Even if all the wires of a wireframe cube are known to be polygon edges, there are ten possible b-reps that can be defined (six 'boxes', three 'tubes', and one 'cube'). The correct interpretation is ambiguous, and only the closed volume of the cube, which satisfies the Euler equations, could be a solid, though it remains ambiguous whether it is a hollow cube (a boundary representation) or a true solid.

Going the other way, beginning with a representation such as "a unit cube, centered at the origin and aligned with the axes," you can easily work out a boundary-representation listing the six faces and their corner coordinates. The boundary-representation, in turn, is easily turned into a wireframe representation

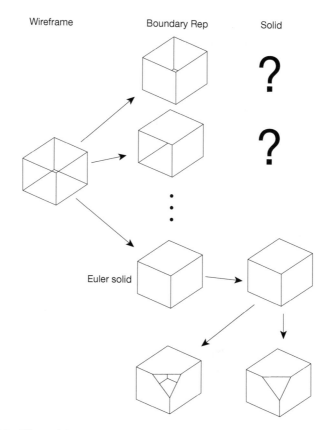

FIGURE 5.5 The ambiguous meaning of wire-frame data.

by turning the edges of the faces into simple lines. At each stage we've used higher-order geometrical knowledge, but discarded some of it in performing the conversion. Wireframe, b-rep, and solid representations contain different amounts of topological and surface information in addition to the coordinates of the vertices. It is easy to discard information; once gone, assumptions or interpretation (guessing) are required to get it back.

The Graphics Pipeline

The term *graphics pipeline* refers to a series of steps needed to convert 3D geometry data into a 2D screen image. While variations exist, and rendering algorithms such as ray-tracing take a very different approach, the pipeline is so standard that it has been implemented in hardware as part of the graphics subsystem of most computers. The complete pipeline, like an assembly line, is composed of several distinct data-processing stages, with data passing along the pipeline unaware of other data ahead or behind. Depending on the desired product, we might leave out steps or add steps to the process.

Just as a building doesn't require a human presence to exist, 3D geometry data need not reference the human eye. However, to draw an *image* or *rendering* of a 3D scene, reference must be made to a human viewer or a camera. Most texts on the subject use the term "eye-space" to describe the world as seen from the camera, with the *x*-axis pointing off to the right, the *y*-axis pointing up to the sky, and the *z*-axis pointing into the scene.

To control the camera or eye, the program will provide a mechanism (called a viewing model) by which the user can adjust position. There are two basic ways to approach this: the environmental viewing model and the object viewing model. While they are equivalent, in that each may be used to specify any view possible in the other, the relative ease with which a given view is specified is very different.

In *environmental* viewing models the user positions the camera or eye point in the same geometrical world (i.e., in the same coordinate system) as the geometry itself. Thus, very precise questions may be asked, like "How much of that building will I see if I stand in the mayor's office and look out the window?" As suggested, this is a fairly common model for architectural visualization.

Object viewing models behave as if the camera is fixed in space and the object (the model) is being manipulated in front of it—as if you held it in your hands. You bring it closer, turn it this way or that, and so on. You always look toward the center of the model but you never know exactly where you are relative to the geometry. This is a common model for engineering part design. It doesn't work well if the point you want to look at isn't near the center of the model (e.g., when generating a pedestrian's view of the street in an area of high-rise buildings).

Not all parts of the model will be visible. The *clipping* process removes the parts that are outside of the view frustum—a pyramidal volume defined by the edges of the screen or window and the eye-point (Figure 5.6). Objects that are not within

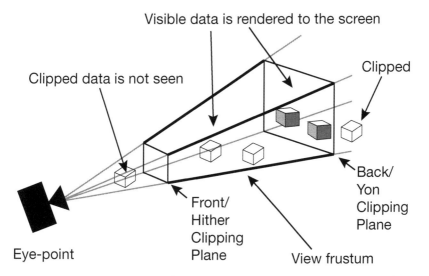

FIGURE 5.6 A 3D view frustum related to clipping by front and back (hither and yon) clipping planes.

this volume are not rendered in the final image. Some software enables a "near/far" or "hither and yon" clipping constraint as well, with planes along the line of sight, effectively enabling perspective section views.

Rendering of the visible data begins with a process called *projection*. Using matrix multiplication, the 3D geometry data is flattened into 2D geometry image data. While many projections exist, most programs offer only *perspective* and *parallel*. You can think of the projection like the lens on a camera—when you point a camera at a scene the type of lens (zoom, wide angle, fish-eye) on the camera changes the character of the image on the film.

Clipping and projection are just the beginning of the process. If the program simply projected the vertexes of the model into the image plane and connected the dots it would create a wireframe graphic in which all lines are visible. To simulate opacity, programs include algorithms for *hidden line* or *hidden surface* removal to make the model look solid by obscuring or not drawing the hidden lines or edges. These algorithms determine which geometry in the model is responsible for particular pixels on the screen, after which they are carefully colored or shaded. Each of these algorithms makes trade-offs between computation time and image quality.

Once projection and hidden surface removal have occurred, the visible surfaces are rendered. This step depends in part on the hardware you are using (you can't display color on a black-and-white screen). Surface color (a data attribute), light-source information (more data), and algorithmic shading models that describe how light interacts with the surfaces of the model all contribute to *shading*. Such algorithms are generally called *flat* or *cosine* shading algorithms. When models have been tessellated (like a hexagonal tube) and we want them to *look* smoothly curved (like a true cylinder), some form of *smoothing*, such as Gouraud (Newman and Sproull 1973) or Phong (Foley *et al.* 1990) is also used.

During shading the opportunity arises to define variable surface coloration or bumps. *Texture mapping* allows these surface characteristics, and others, to be applied to the geometry like wrapping paper or wallpaper.

Separate from shading is the casting of *shadows*. Shadows occur in those parts of the scene that you can see, but the sun or other light source cannot. One approach projects the model's edges, as seen by the sun, through space to define shadow volumes. Another computes a view of the scene from the sun's point of view and then projects that as a texture onto the elements of the model. For close-in views of large models or low-resolution shadow maps, the result is often distinctly jagged.

The various steps of the graphics pipeline algorithms have been turned into hardware on graphics processing units (GPUs), the hardware that drives rendering for video games, immersive worlds, and interactive 3D modeling programs.

More complicated rendering algorithms with names like *ray casting*, *photon-mapping*, and *radiosity* take into account how light bounces between surfaces within the model, but take longer to run. With moderately complex data they can take several minutes to render an image, even on a fast computer. They also make images that look so real that they are called photorealistic.

In an interesting convergence, rendering algorithms such as radiosity make realistic-looking images by employing what we know about the physics of light. In contrast to many renderings in which extra light sources and textures are added or shadows suppressed to adjust the appearance of the scene, such physically based renderings are not renderings so much as simulations of the scene. That is, the results are calculated to be correct, not just look nice.

This pursuit of rendering as reality simulation is deeply rooted in the computer graphics industry, but there is recognition that it has ignored rendering as emotional communication or visual exploration. There is room to investigate the communications and decision-support roles of images—still, animated, and immersive—in understanding and memory of virtual spaces, as well as the potential of non-photo-real (NPR) rendering to communicate or visualize other aspects of our environment (heat-loss, air-flow, occupant circulation, glare, etc.).

Applications Across the Building Lifecycle

The major categories of software used routinely in the AEC industry include office automation, file sharing and project management, 2D and 3D sketching, 2D and 3D CAD, BIM, clash-detection, scheduling, and cost-estimating.

Office Automation

While not the differentiating use in the AEC marketplace, there is no question that tools for word processing, spreadsheets, file sharing and file management, operating on top of networks of high-performance personal computers (PCs) and file servers are part of almost every office. Businesses with multiple offices are often linked by networks and shared file systems. Increasingly, firms have added cloud-computing resources for computation-intensive activities, and outsourced selected tasks such as high-end animation or rendering.

Due to the potential costs of mistakes or delays in the construction process, the desire to document and control information flow and access between parties to the process has prompted development of formal systems of information distribution and archiving.

Sketching

Sketching refers to casual 2D and 3D data constructs that are almost always seen as preliminary or temporary documents. Simplified interfaces enable some of these programs to be learned quite quickly, in contrast to the more complex drawing production systems which can take months to master. The broad availability of 3D graphics hardware in PCs means fairly simple models can be quickly generated and manipulated during site visits, client meetings, and design sessions, and used to produce simple textured and shaded renderings. Two-dimensional sketching, possibly on touch-displays, may include line-weight variation in the sketch, and copy-and-paste entourage elements, but may not extend to such document-centric features as dimensions, schedules, etc. While there are usually data-transfer paths by which models may be moved to drawing production systems, re-construction is often easier than import-and-repair, as imported sketch data is often not accurate enough to use going forward.

2D CAD

Two-dimensional CAD systems are ubiquitous in architectural practice. At the heart of CAD software is a vector-based drawing. Each drawing file can have only one editor at a time, each usually represents a single sheet or layer in the documents, and each is a simple list of primitives (lines, circles, text) whose visibility and appearance is governed by local or shared (layer) attributes. Invariant graphic elements (blocks or symbols) may be imported or defined and then placed repeatedly in the drawing, or saved as separate files.

Since they first appeared in the 1980s, these systems have grown very elaborate, with networks (including cloud storage) enabling sharing of passive (read-only background) data, and a combination of usage strategies and software features supporting complex drawing sets at varying scales, with a minimum of duplication. Dynamic *parametric* graphic elements such as other design documents, elements that change appearance at different scales, and elements that change appearance depending on the use case, are all becoming more common. Challenges regarding scale and size of graphic elements such as labels, dimensions, etc. have been addressed by differentiating between paper space (for plotting) and model space (for project data). Live-link inclusion features (*reference* or *xref*) facilitate change propagation by linking files together, helping to reduce replication, though the resulting tangle of interconnected files can make project management and milestone archiving a challenge.

3D Modeling

While there are some offices that limit their use to 2D, most CAD software is actually 3D and capable of some degree of rendering beyond basic wireframe display. This means that 3D architectural CAD can be seen, largely, through the same lens as 2D CAD in terms of tools, structure, and trends. However, there are some differences, including the addition of sophisticated rendering functionality and the emergence of building information modeling (BIM, see next section).

Many disciplines use 3D data, including architecture, visual effects, animation, and engineering. Each of these requires slightly different features in their 3D modeling software, such as rendering, kinematics, motion-control, and digital fabrication. At this time, someone wishing to perform all of these tasks in a workflow would need either a collection of add-ons for their 3D modeler or several different tools, and it is likely they would have to build more than one shared model.

Thanks in large part to research done in support of the motion-picture industry, and the subsequent rise in expectations in general, powerful and accurate 3D rendering with realistic light-transport, surface models, textures, and synthetic camera controls is now available on PCs, though heavy use still requires multi-PC render-farms or cloud-computing resources to bring rendering times down from days to hours.

Building Information Modeling (BIM)

Around the year 2000 the concept of a "building model" began to catch hold. The concept originated as far back as the 1970s, but it depends on the use of a centralized data system such as that found on mainframe computers. When PC systems (which were not networked, had limited storage, and were not very fast) became popular in the 1980s for economic reasons, users were forced to accept the "each file is a drawing" paradigm as well. Now that PCs have multi-user operating systems, fast networking, and much more powerful hardware, the idea has re-emerged as BIM. In contrast to 2D CAD, in which the data represents conventional working drawings, and 3D modeling, in which the data represents only geometry and surface properties, a BIM represents a virtual building in 3D. The primitives are walls and windows, not lines and arcs or polygons. They have material properties in addition to geometry. Multiple users may edit the model at the same time. Walls can be rendered differently at different scales. Software extracts 2D drawings and schedules from the central model, assuring consistency within the construction document set by minimizing repeated representation of features that appear in more than one drawing. In addition to geometry, BIM data may include material or manufacturer information, analysis results, and links to related information such as operating manuals. This allows a BIM to serve as an information repository throughout the life of the building.

BIM is more demanding computationally than 2D or 3D CAD. Modern high-end PCs and networks can usually deliver the necessary communications and computation, but Martyn Day reports (Day 2011) that Autodesk had to use Navisworks software to do clash detection in the redesign of their Boston offices, because the BIM model was too large for Revit to handle the task on available hardware.

BIM is widely seen as critical to the effective future use of computing in the AEC market and is linked to the emergence of alternative practice forms, such as integrated project delivery (IPD), that attempt to enhance construction outcomes by sharing data and expertise more efficiently throughout the design and construction process. As a strategy, IPD aims to introduce appropriate expertise into project development at a time when the impact is greatest. It is often accomplished through sharing of a BIM dataset. As responsibility for creating the data and benefits from use of it shift between the different players in the process, it is likely that further change will occur (Scheer 2014).

Another force for adoption and further enhancement of BIM technology comes from large institutional and government owners, operators, and regulators of buildings, who have begun to perform more systematic cost and/or energy analysis of designs in their oversight roles, as part of which there is a clear trend to push or require use of BIM tools at some stages in the design process.

The move to BIM faces a number of challenges, including interoperability, intellectual property rights, incorporation of consultant expertise, mismatches of model granularity, representational range, change management, and phasing (use from sketching to construction and occupancy). We'll come back to these subjects under the general headings of "Representation" and "Expertise" in the next part of the book.

Analysis, Simulation, Optimization, and Automation

Although document production via BIM or CAD is important in a practice, it is not the only area of digital tool use. Software for structural, energy, air-flow, and lighting analysis has grown very sophisticated in the last decade, with easier-to-use interfaces for less expert users, and with (somewhat) automatic conversion from standard 3D model and BIM file formats, while producing better-looking graphic results. Production of analysis results has gotten easier but, by and large, identification of the changes needed to improve the design remains the designer's task.

The exception to this may reside in the rapidly evolving domain of scripting and the related field of parametric design, in which practitioners, researchers, and academics are using automated analysis tools to produce assessments and creating feedback loops linked to generative design strategies such as genetic algorithms and parametric optimization. We'll talk about these in more depth in Chapter 12.

Fabrication and Construction

Construction is a collaborative activity; the work of many individuals needs to be coordinated in space and time to complete the project in a safe, economical fashion. Even before construction begins, the designers consult with specialists regarding aspects of the project. The representation created by the designer or design team will almost certainly become the central means of communicating design intent to other members of the team. The representation must fit the cultural frame, as well as the cognitive and information needs of the designer, the consultants, and the contractors.

Contractors make extensive use of software for tracking (and costing) material, scheduling material delivery and work crews, and detecting clashes or interference between building subsystems prior to construction.

While computer numerically controlled (CNC) machines have been around for several decades, they have until recently seen limited application in construction. One reason might be that the size of a fabrication tool is usually greater than the part being made, making buildings difficult to fabricate at full size. The aesthetics, analysis, and economics of the modernist box all favored use of identical repetitive elements. However, our computer systems and parametric design software now enable us to work quickly and reliably with unique sizes and dimensions rather than representative units, while digital manufacturing and scheduling systems have made it possible to manage the analysis, production and delivery of these many one-off design elements at modest premiums, making custom-cut steel sections and custom-fabricated façade elements almost commonplace in many situations where corporate branding or technical challenges warrant.

Operation

Building owners and operators need information about their buildings. In addition to knowing operational details such as which tenants pay for what offices or services, or who to call if there is a break-in or a leak, they may need to know what equipment is installed, how to operate it, who provided it, what maintenance operations are required, and when they were last performed. Over time, as building use changes, they are likely to need to know where conduit and plumbing is hidden in walls, which data ports are live, which electrical circuits serve what areas of the building from which panels, etc. They don't need everything the designer or contractor needs, but they need a substantial part of it.

In the past this information would be provided in the form of an *operating manual* containing copies of the construction documents and all the equipment manuals. It was then up to the operator to maintain and update that information. Acquiring them in digital form seems ideal, as they can be extended with post-occupancy data, backed up regularly, and, on multi-building campuses, maintained by a central management team. A BIM may also be integrated into the building's real-time *building automation system* (BAS).

Building operations rely increasingly on real-time systems for lighting, access-control and security, elevator operation, control of automated shading and irrigation systems, heating, ventilation, and air conditioning (HVAC) system operation, plus the more esoteric jobs of monitoring photovoltaic (PV) array output and managing gray-water recycling. Selection of these systems occurs during design, but owners are responsible for operating, maintaining, and possibly updating them. Commissioning of a building often includes handover of a BIM model corresponding to the as-built conditions, with digital versions of manuals, invoices, maintenance schedules, etc. While there is concern about maintaining such repositories over time (Johnson 2014), the information gives owners a strong foundation for implementing a BAS.

Increasingly, design of the building needs to include operational considerations and might involve selection or specification of particular building or equipment features and their integration into the BAS. Complex buildings may require complex management teams, but they also challenge designers to consider operation during design. As much of this will involve digital technologies, it fits within the umbrella of design computing.

Summary

Offices have built complete project workflows around 2D CAD using complex hierarchies of file-in-file referencing to minimize duplication while allowing information to appear at different scales and in different contexts. Some sketching (2D or 3D) is occurring, though such models may be re-constructed during production. Three-dimensional BIM is rapidly replacing 2D CAD for design development and construction documents, though it is often thought to be too cumbersome to use in schematic design and there remain issues regarding data interoperability when it comes to sharing design data with owners, contractors, and consultants. Increasing use is being made of all-digital file-to-factory fabrication workflows, including weatherproof modular off-site construction, though this is not yet widespread. Building owners such as the government and large corporations are pushing BIM in the belief that it will reduce costs and improve the product they receive and that it will make ongoing management more productive, though only the technologically more advanced building operators may see long-term benefits.

Suggested Reading

Fallon, Kristine. 1998. Early computer graphics developments in the architecture, engineering and construction industry. *IEEE Annals of the History of Computing* 20 (2): 20–29.

Foley, James, Andries van Dam, Steven Feiner, and John Hughes. 1990. *Computer graphics principles and practice* (2nd edn). Reading, MA: Addison-Wesley.

Scheer, David Ross. 2014. *The death of drawing: Architecture in the age of simulation*. New York, NY: Routledge.

References

Day, Martyn. 2011. The trouble with BIM. *AECMagazine* (October). http://aecmag.com/technology-mainmenu-35/450-the-trouble-with-bim.

Fallon, Kristine. 1998. Early computer graphics developments in the architecture, engineering and construction industry. *IEEE Annals of the History of Computing* 20 (2): 20–29.

Foley, James, Andries van Dam, Steven Feiner, and John Hughes. 1990. *Computer graphics principles and practice* (2nd edition). Reading, MA: Addison-Wesley.

Henderson, L. R. and A. M. Mumford. 1990. *The computer graphics metafile*. Borough Green: Butterworth.

Johnson, Brian R. 2014. One BIM to rule them all: Myth or future reality?, in *Building information modeling: BIM in current and future practice*. Edited by K. Kensek and D. Noble, 175–185. Hoboken, NJ: John Wiley & Sons.

Newman, William and Robert Sproull. 1973. *Principles of interactive computer graphics*. New York, NY: McGraw-Hill.

NIBS. 2014. *United States National CAD Standard: V6*. Washington, DC: National Institute of Building Sciences. www.nationalcadstandard.org/ncs6/history.php

Scheer, David Ross. 2014. *The death of drawing: Architecture in the age of simulation*. New York, NY: Routledge.
Sutherland, Ivan. 1963. SketchPad: A man–machine graphical communication system. *AFIPS Conference Proceedings* 23: 323–328.

PART II
The Grand Challenges

With some understanding of how the world of design looks in the mirror of computation, and acquainted with the current state of the art in architecture, engineering, and construction (AEC) information technology, we can consider the core challenges, questions, or puzzles that brought us here and continue to motivate, frame, and define the emergent field of design computing. There is no reason to see this as a definitive list, nor are these particular topics or questions the only way to organize the pieces, as it is difficult to pull out distinct topics because they are interdependent. For example, the question of design problems necessarily touches on solutions, representation, cognition, and interaction, but it also stands on its own. In most cases, the component issues predate the influence of computing, so most questions have a history of theoretical development, but each has been challenged, sharpened, or illuminated when viewed through the lens of design computing.

The order chosen here allows earlier discussions to be referenced in later chapters but is otherwise not special.

Problems—What are design problems? How are they different from known and routine problems? How do we understand problems? Symbolic processing divides problems into parts; the notion of state spaces and design as search of state space. Situated action acknowledges social aspects, wicked problems, and the co-evolution of problem statements and solutions. Can both be right?

Design Cognition—How do designers think? What is design process? What is the role of drawing, memory, divergent and convergent thinking, learning and experience? How are ambiguity and emergence involved? How can inspiration and intuition interact with analytical and systematic attempts to capture knowledge from design process and the built world?.

Representation—What role do representations play in design cognition? How do representations interact with tasks? Challenges in representation: What makes the single model difficult? How can we allow emergence, or show ambiguity or certainty in a design? How do different users and uses influence choice of representation?

Interface—Creating powerful interfaces without overwhelming complexity. Accommodating different users and needs; hand gestures and careful pointing. Bridging digital and physical modalities. The challenge of the neutral tool: affordances, desirable difficulty, and seduction.

Practice—The productivity challenge. Information technology is creating opportunity, shifting roles, legal relationships, internal power, and status across the AEC industry. Information access, creation, management, and monetization across projects and time is increasingly important.

Expertise—Design is increasingly about coordination of multiple consultants. How do we tap into expert knowledge? How is their view of the design model different? How can designers incorporate results sooner and with more impact without increasing cost?

Solutions—Searching design space involves: generation, testing, stopping. Are there ways to find solutions algorithmically—to make design computable? Are there ways to empower human designers to find solutions more quickly? What is the best way to move from vague and ambiguous to well defined and precise?

Buildings—What is a smart building? What opportunities exist to enhance buildings through application of ICT? What does an architect need to know to design a smarter building? How might big data change design processes and buildings?

Pedagogy—Are younger generations different? How do we train young architects to do design? Does computer use damage creativity? Can we re-train mid-career designers? What is the relationship between computing skills and continuing education expectations?

6

DESIGN PROBLEMS

What Are They?

> Everyone designs who devises courses of action aimed at changing existing situations into preferred ones.
>
> *Herbert Simon (1969)*

The nature of problems in general, and design problems in particular, has challenged philosophers, computer scientists, and designers alike. Simon's definition of design as a problem-solving behavior aimed "at changing existing situations into preferred ones" is a common starting point (Simon 1969; Jackman *et al.* 2007). Everyone does it. It is a cognitive behavior. The goal is an improvement. Some "existing situations" are problems; but are all problems design problems? If I need my hat off the shelf in the closet and I simply stand up and reach for it, I may be changing an existing situation to a preferred one, but have I done design? The solution to the problem is immediately apparent—*obvious*. It simply needs to be executed. Only if I am unable to reach the hat do I need to think about it. Even then, having encountered this difficulty before, I may know I can stand on a chair and reach, or use a stick to work the hat off the shelf. I may know of several alternative solutions requiring a step or two to execute (fetching the chair or finding a stick), but the solution is *routine*. Only someone contemplating this problem for the first time actually needs to devise a new, or creative, response. In each case the existing situation is the same, as is the preferred one, indicating that response to problems depends on personal experience and that the response will not always be unique or novel.

Simon's "course of action" draws attention to the *process* of retrieving the hat, but the desired result (hat on head) is also a physical configuration of the world that addresses the problem of a wet, cold head. Thinking to put on a hat involves *knowledge* that the hat is an appropriate solution, one among several. Both process and knowledge are involved in design (Kalay *et al.* 1990).

Defining the Problem

> First, the taking in of scattered particulars under one Idea, so that everyone understands what is being talked about.... Second, the separation of the Idea into parts, by dividing it at the joints, as nature directs, not breaking any limb in half as a bad carver might.
>
> *Plato*

In his 1964 *Notes on the Synthesis of Form*, Christopher Alexander uses this quote from *Phaedrus*, in which Plato talks about "the taking in of scattered particulars under one Idea, so that everyone understands what is being talked about," acknowledging the difficulty of establishing the scope or boundaries of many problems (Alexander 1964). Alexander uses the problem of producing hot water for tea to illustrate the range of solutions, from kettle to household to municipal hot water possibilities. Locating, bounding, and defining the *Idea* that identifies the problem remains a challenge.

Having established an Idea, Plato's second step is "the separation of the Idea into parts, by dividing it at the joints, as nature directs, not breaking any limb in half as a bad carver might" (Alexander 1964). The notion of problems being divisible into parts is crucial to the theory of classical problem solving proposed by Newell and Simon (1958) because it is part of divide-and-conquer strategies which attempt to reduce complex problems to solvable routine parts whose individual solutions can be aggregated into a complete solution.

The "taking in of scattered particulars under one Idea" and "separation of the Idea into parts" also suggests that all characteristics of the problem can be established before the solution process begins. This view sees both the problem and the solution plan as objective facts in the world, distinct and separate from us, the designers. Unfortunately, many problems are impossible to separate from their context; they are unavailable for laboratory study and their "joints" are difficult to define in advance, producing what Rittel and Webber (1973) describe as *wicked problems*. This observation that problems depend on social context has also seen significant exploration in the computer science field of human–computer interaction (HCI), where the theory of *situated action* has gained significant currency (Suchman 1987).

Researchers have also found that the perceived difficulty of a problem is not an intrinsic quality of the problem; it tends to depend on the level of expertise of the designer (Dorst 2004). What a neophyte might find a challenging design problem, an experienced designer may see as routine, or even obvious. The lack of fixity to problems makes it difficult to develop software that is universally appropriate—two different designers may need very different kinds of tools, or a single designer may need different tools at different times.

If the action required to transform the existing situation to the preferred one is recognized at the same time as the problem, the solution can be said to be obvious or known. This may happen because the exact same problem, or a very similar problem, has been encountered in the past and its solution is remembered.

More challenging than problems with obvious solutions are problems that might be described as nearly obvious or routine—those with well-known procedural solutions. "Routine design can be defined as that class of design activity when all the variables which define the structure and behavior are known in advance, as are all the processes needed to find appropriate values for those variables" (Gero 1994). Many engineering problems, such as the selection of a beam to carry a particular load, are well understood and characterized by mathematical formulas that can be solved in a straightforward fashion, given knowledge of appropriate material properties.

Where no obvious solution is available, where analogy to similar problems fails, and where no analytic formulation exists, designers deploy a process or set of interrelated strategies intended to help identify solutions. While descriptions of design process vary, one of the first steps will probably be to analyze the problem and identify its salient characteristics. In architecture this may take the form of a survey of required spaces, identifying the uses and required size of spaces, adjacency requirements, client's style preferences, etc. The process of problem definition, dividing the problem "at the joints," begins to establish the problem solution: If you need a house, an office building isn't the solution.

Design Spaces

Problems and designs are usually understood in terms of the set of *characteristics* that are sought (cozy, modern, energy efficient, etc.) and *measures* (cost, sizes, number of bedrooms, number of users, annual energy budget, etc.) that must be achieved. It sometimes helps to think of each of these characteristics and measures as separate distinct dimensions, perhaps with some known values (e.g. maximum cost), but with many unknowns. Collectively, the characteristics and measures define a multidimensional *design space* or *state space* within which the ultimate solution must be sought. It is the set of all unique design solutions that can be described by the selected characteristics. A design proposal can be assessed along each of the axes of this space (count the bedrooms, estimate the costs, simulate the heat loss, etc.). The values establish a *state* in that space (a point or a region consisting of closely related points). This is an important idea that appears repeatedly in this book and much related literature.

One of the first modern descriptions of design space and state space search came from Herbert Simon, who won the 1979 Nobel Prize in economics. In his 1969 book, *The Sciences of the Artificial*, he presented his ideas about how to study the man-made, artificial world and its production. A powerful concept, design space remains one of the backbone ideas of design computing.

Implicit in this idea is another—that design is search. This idea is consistent with observations of both paper-based and digital design activity (Woodbury and Burrow 2006; Jackman et al. 2007). It underpins many projects that aim to support or augment human design skills, and matches the description of design as iteration. The biggest problem, as we will see later, is the enormous number of candidate states.

State space search strategies rely on measurable aspects of the design, but not everything you value about an environment can be measured. There may be affective qualities such as "relaxed" or "corporate" which are important, but not directly represented in the design, nor computable from what is represented. That is, no matter how detailed, state space representations are always incomplete approximations of the overall project.

Fitness or Objective Functions

Closely associated with the idea of a design space is the idea of an objective function, also called a *utility* or *fitness* function, discussed in more depth in Chapter 12. For each design this function combines both the measures along the design space dimensions and a weighting based on the importance or impact of each measure to produce a value that represents an assessment of the design. This value provides a means by which designs may be compared.

If you think of the face of the earth as a two-dimensional design space of locations, the elevation of the surface at that location might be thought of as the objective function. The problem of finding a high or low spot is a simple optimization problem.

Constraints and Goals

During problem definition or exploration, the characteristics that the design needs to satisfy might be expressed as constraints ("five bedrooms") or performance goals ("net-zero energy"). As part of an objective function these can be thought of as weights applied to measures along the axes. However, while we may think of the separate axes of the design space as distinct, it is often a mistake to think of them as orthogonal, or independent. For instance, the volume of heated space in a building impacts the energy consumption, so the number of bedrooms affects the ability to get to net-zero energy. One of the challenges and wonders of design is the ways in which multiple design goals can sometimes be met by a single solution, as when windows provide both views out and daylight in. Of course, it is also common for constraints and goals to conflict, as happens with "minimize cost" and "maximize space," precisely because they are interrelated. Because nominally different constraints or goals actually interact and depend on the value system of those making assessments, discovering the core challenges of a design problem can be quite difficult, as Rittel and Webber discussed. Design problems with too many criteria may be overconstrained (impossible to solve), while those with too few will be *underconstrained* (with too many solutions).

Puzzle Making

There is another view of design problems that sees them co-evolving with their solutions (Dorst 2004). This view acknowledges the human difficulties of fully defining the problem up front; sometimes it is helpful to redefine the problem. The particular solution and the problem definition emerge together from a process of *puzzle making* (Archea 1987).

The view of design as puzzle making or co-evolution is very different from the divide, conquer, and aggregate models descended from the Platonic view. It acknowledges the active role of the designer in formulating the problem. While Newell and Simon's work was intended to lead to artificial intelligence, puzzle making leaves human cognition at the center of the process. It also suggests that wicked problems may arise from a mis-fit of problem and solution, requiring renegotiation of either or both.

The Importance of Problem Definition

Identifying the true problem motivating a design project is not always easy. One, possibly apocryphal, story involves a growing company contemplating construction of a new building. They are said to have solved their space needs by switching to a "hot desk" model of space assignment after their architect pointed out that the number of workers traveling on any given day was about the same as the number meant to be housed in the new building. By assigning open desks to employees on an as-needed basis, they saved the cost of a new building.

This may all sound esoteric and obscure, but it was mis-alignment between problem definitions and proposed building designs that helped motivate the US General Services Administration (GSA) to promote BIM technology for design, beginning in 2003 (Hagen 2009). The GSA specifies and justifies buildings based on spatial needs, both in terms of areas and circulation (many federal buildings are courthouses with intricate public, judicial, and defendant circulation systems), but proposals and plans submitted by architects were difficult to compare against the specifications. The first-phase BIM deliverable requirement for GSA projects beginning in 2007 was for a *Spatial Program BIM*, effectively requiring those aspects of a design to be interpretable by computer in a standardized way.

The Problem of Solutions

Generation and refinement of solutions within the design space is the subject of Chapter 12, but two points are important to note here. The first is that the way in which the problem is defined or described—Simon's "existing conditions" and their "preferred" alternatives—goes a long way to determining the shape of the solution by establishing the vocabulary. This observation reinforces the importance of careful problem specification, especially if automated tools are to be used.

The second and related challenge of solutions to problems is "the stopping problem." To be useful, an automated process should be able to recognize when a solution has been reached. Each iteration of the design is different, but not necessarily better, and there is no way to identify an objectively "best." In the absence of a good stopping rule the design process may continue until some external event (budget or calendar) draws it to a close. Defining problems in such a way that a solution can be recognized when it emerges remains a grand challenge.

Summary

Design is a problem-solving behavior. Distinctions can be made between obvious, routine, and creative design problems, the latter requiring action of sufficient complexity that planning is necessary. Design expertise influences the perceived difficulty of the design problem. While it is generally asserted that problems can be divided into their natural parts during analysis or during the solution process, research into wicked and situated problems supports the notion of puzzle making or co-evolution of problem definitions and solutions. Characteristics by which the problem is defined become characteristics by which a solution is measured and constraints defined. These can be taken to define a design state space that is useful for conceptualizing the problem–solution pair. This space provides a domain over which a utility or objective function might be defined, providing a framework within which to assess solution quality and perform design iterations until a suitable design is recognized or a stopping rule satisfied.

Suggested Reading

Alexander, Christopher. 1964. *Notes on the synthesis of form*. Cambridge, MA: Harvard University Press.
Dorst, Kees. 2004. The problem of design problems, in *Proceedings of design thinking research symposium* 6. http://research.it.uts.edu.au/creative/design/index.htm.
Kalay, Yehuda, L. Swerdloff, and B. Majkowski. 1990. Process and knowledge in design computation. *Journal of Architectural Education* 43: 47–53.
Rittel, Horst and Melvin M. Webber. 1973. Dilemmas in a general theory of planning. *Policy Sciences* 4: 155–169.
Simon, Herbert. 1969. *The sciences of the artificial*. Cambridge, MA: MIT Press.
Suchman, Lucy. 1987. *Plans and situated actions: The problem of human–machine communication*. Cambridge: Cambridge University Press.

References

Alexander, Christopher. 1964. *Notes on the synthesis of form*. Cambridge, MA: Harvard University Press.
Archea, John. 1987. Puzzle-making: What architects do when no one is looking, in *Computability of Design*. Edited by Y. Kalay, 37–52. New York, NY: John Wiley.
Dorst, Kees. 2004. The problem of design problems, in *Proceedings of design thinking research symposium* 6. http://research.it.uts.edu.au/creative/design/index.htm.
Gero, John. 1994. Introduction: Creativity and cognition, in *Artificial intelligence and creativity: An interdisciplinary approach*. Edited by T. Dartnall, 259–268. Dordrecht: Kluwer.
Hagen, Stephen, Peggy Ho, and Charles Matta. 2009. BIM: The GSA story. *Journal of Building Information Modeling* (Spring): 28–29

Jackman, J., S. Ryan, S. Olafsson, and V. Dark. 2007. Meta-problem spaces and problem structure. In *Learning to solve complex, scientific problems*. Edited by D. H. Jonassen. Mahwa, NJ: Lawrence Erlbaum.

Kalay, Yehuda, L. Swerdloff, and B. Majkowski. 1990. Process and knowledge in design computation. *Journal of Architectural Education* 43: 47–53.

Newell, Alan and H. A. Simon. 1958. Elements of a theory of human problem solving. *Psychological Review* 65: 151–166.

Rittel, Horst and Melvin M. Webber. 1973. Dilemmas in a general theory of planning. *Policy Sciences* 4: 155–169.

Simon, Herbert. 1969. *The sciences of the artificial*. Cambridge, MA: MIT Press.

Suchman, Lucy. 1987. *Plans and situated actions: The problem of human–machine communication*. Cambridge: Cambridge University Press.

Woodbury, Robert and Andrew L. Burrow. 2006. Whither design space? *AIE EDAM: Artificial Intelligence for Engineering Design, Analysis, and Manufacturing* 20: 63–82.

7

COGNITION

How Do Designers Think?

> Inspiration is for amateurs. The rest of us just show up and get to work. If you wait around for the clouds to part and a bolt of lightning to strike you in the brain, you are not going to make an awful lot of work. All the best ideas come out of the process; they come out of the work itself.
>
> Chuck Close (Close 2002)

Observing what designers *do* has helped produce many powerful software tools, but the question of how designers *think* is important because the traditional pre-digital tools we can observe being used in traditional practice might not, themselves, offer the best support for design. The question of design cognition has been addressed by numerous authors and disciplines over the years, including computer scientists Newell *et al.* (1958), designers Cross (2011), Lawson (1997), and Schön (1984), design computing researchers Akin *et al.* (1987) and Gero (1990), and psychologists Kounios and Beeman (2009). Touching, as it does, on human creativity, insight, and inspiration, the answer has proven elusive but worth exploring.

The epigraph touches on two common but seemingly opposed views of how design occurs: an "Aha!" bolt-of-lightning moment of insight (Kounios and Beeman 2009), or a reward for persistent effort (Close 2002). A careful reader might note that Close contrasts persistent work with "wait[ing] around" and finds that ideas "come out of the work," while Kounios and Beeman acknowledge "studies show that insight is the culmination of a series of brain states and processes operating at different time scales" (Kounios and Beeman 2009). Rather than being opposing views, it seems possible that the two views describe aspects of a single process in which large- and small-scale insights grow out of close attention and engagement with a complex problem. The challenge is to capture and organize the insights while fostering attention and informing engagement.

One fundamental cognitive problem is that products such as buildings are complex, and yet the working or short-term memory of a human holds relatively little information. Long-term memory holds much more but takes longer to access and cannot be used for a dynamic process such as design. External representations, such as drawings, aid in the process, as do models that divide problems into hierarchies of smaller conceptual fragments or chunks that can be manipulated within the limits of short-term memory. Design concepts usually engage the problem by establishing relationships between a relatively small number of elements, creating a theme that can be used to address challenges at different scales and in different aspects of the design. Good concepts are not arbitrary; they fit the design problem. They "come out of the work itself" (Close 2002) and they "are often the result of the reorganization or restructuring of the elements of a situation or problem" which produces "a new interpretation of a situation ... that can point to the solution to a problem" (Kounios and Beeman 2009). Before you reorganize a situation, you have to study and understand it; you must build a (conceptual) model of it. This means identifying, labeling, and organizing parts (building a representation), and defining changes or "moves" that can be carried out on the parts. If a subsequent insight or design concept re-organizes the elements of the problem, you may need a new model.

Design, whether insightful or sweat-stained, does not require a computer. Still, as information technologies have been successfully applied to observable productive design behaviors (typing, drafting, modeling), we are moving towards supporting or automating the less observable cognitive aspects of the activity, where architects' professional identities and egos are at risk. That risk arises not only from the core difficulties of defining any design profession, but also from the notion of making design tasks computable. Nonetheless, beginning in the mid-1990s, conceptual models and a series of conferences have wrestled with aspects of creativity, design, and cognition (Gero 2011). In asking how designers think, we not only open the door to significant productivity gains for designers, we open the door on the computation and possible automation of design—replacing designers.

Whether we aim to automate design, enhance designer productivity, or just support one aspect of design, such as collaborative interaction, we need to develop a model of the process. Out of this may emerge points of leverage where computing might be well applied, or hesitancy, where (current) computing technology may be more hindrance than aid. We must look at the design process, at divergent thinking and the role of memory, media, and social process, and at convergent thinking leading to incremental improvement. In addition, as digital media increasingly take center stage in offices, we need to consider how their use interacts with our thinking and the possibility that their use has unanticipated side-effects.

Designing as Process

> Invention is "one percent inspiration, ninety-nine percent perspiration."
> *Attributed to Thomas Edison*

One view of design and designers tends to focus on the *idea* itself, the designer's proposition about what should be done, not where the idea came from. Consequently, there is a certain amount of mysticism regarding the process. People are said to "be creative," not "act creatively" and we marvel at examples of "thinking outside the box" without necessarily being aware of or appreciating the amount of work that preceded the "Aha!" moment. Production of the idea is then sometimes seen as an act of will and design as a domain of willfulness. However, while professional designers respect the possibility of sudden inspiration, it is more often seen as a fickle source of design solutions.

The general consensus is that design is a behavior that follows a fairly common pattern that can be taught and cultivated, and that certain techniques (e.g., precedent studies, brainstorming, change of perspective) enhance creativity. In *The Design of Everyday Things*, Don Norman presents a very general model for the "seven stages of human *action*" that aligns with the stages of design shown in Figure 7.1 remarkably well (Norman 1988). His model begins with a *goal* (the design problem, or "preferred state"), formulates an *intention* of a change ("analysis"), and offers a specific plan of *action* meant to achieve it ("ideation"). The plan is *executed* ("representation"), after which one must "perceive the state of the world" ("simulation"), *interpret* the state of the world ("evaluation"), and *evaluate* the outcome ("assessment"). If not completely successful, a new goal is set and the cycle repeats. This model reminds us that the effect of a change is not always knowable in advance, and that feedback loops containing perception, interpretation, and evaluation form important steps in any action, including design.

Even if design were entirely dependent on a flash of inspiration and there is no way to promote it, there is a role for digital tools, because it will still be necessary to record, evaluate, refine, and document the idea in an external medium such as that provided by a 3D modeling environment or sketching program. While creative production remains a crucial element in design, there are a number of related cognitive activities that are also candidates for support, including problem exploration, solution representation, solution evaluation, decision support, and recording, as we will see in later chapters.

Creativity represents only one aspect of design cognition or what goes on in someone's head during design problem solving. Other aspects include problem formulation, mental representation, and internal manipulation. Depending on what we think the core of design consists of, software augmentation or replacement will take different forms. In the following sections we'll look at design as process or method, intuition or inspiration, memory, selection, drawing or dialog, social process, and as the end product of incremental improvement. None captures all of design; each has been the starting point for software projects in the past.

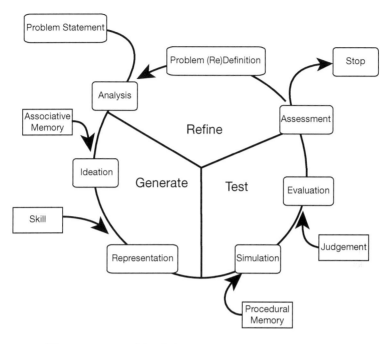

FIGURE 7.1 The seven steps of the design process.

As mentioned in Chapters 1 and 6, and shown in Figure 7.1, design is usually described as an iterative or cyclic *process* that gradually, though not directly, proceeds from problem statement to solution. Beyond that foundation, differences appear. Some distinguish between strong design procedures or methods and strong design knowledge (Kalay *et al*. 1990). Others stress the role of social interaction among the participants, or the gradual construction of both problem and solution, described by more than one author as *puzzle making* or *coevolution* (Rittel and Webber 1973; Archea 1987).

Most models of design cognition identify two kinds of thinking, described in the "brainstorming" technique as *divergent* and *convergent* thinking (Osborn 1953). During divergent thinking (also called associative thinking), the designer is actively collecting as many disparate solutions as possible, without regard for their practicality or details. This is usually seen as the more intuitive and creative aspect of design. During convergent thinking the designer is refining the design idea and considering the ramifications. Divergent thinking is often creative "shower thinking" (Bryce 2014), while convergent thinking is sharp-pencil, "working it out at the drawing board" thinking. The premise in the literature on creativity, as well as remarks by Chuck Close, Thomas Edison, and others, is that process and practice are the primary ingredients of creative productivity.

While being creative may be something one can practice, mastering creative production, as with other expertise, requires time and experience. Experts view problems more abstractly than beginners, so one aspect of developing expertise probably consists of learning templates, frames, or categories of action that can be applied to multiple problems. Young architects are often advised and inclined to travel, in part because it provides grist for the mill of design thinking by associating their own human experience with as many examples of the built environment as possible. They are also advised to study different buildings and cultures, and to record their thoughts in sketchbooks and drawings. Done regularly, the habit builds both direct personal memory and documentary resources.

Design Moves

When considering design computing strategies to provide cognitive support to designers, we need to consider each stage of design, not just one, such as drawing. The creative practice of design has deep roots in the designer's lived experience, and their mastery of the design process in a way that combines divergent and convergent thinking, but it also requires the designer to evaluate alternative designs and devise transformations—what Donald Schön calls "moves"—that can be applied to make the design better (Schön 1984).

Design moves may be subtle or bold. They very often nest, one inside another, as when an entire floor plan is flipped or rotated. Much of learning to design, as Schön discusses, is concerned with developing a repertoire of moves and judgment regarding their appropriate use. Constraints on available moves, through technological limits or designer inhibition, restrict design exploration. Undoing a move, *backtracking*, is common. Making increasingly detailed moves following a hierarchy of importance is generally thought to be efficient.

Unfortunately, the list of possible moves seems to arise from the design problem itself, making them difficult to identify or catalog in advance and giving rise to what Schön (1984) characterizes as "reflection in action." Expert designers also apply abstractions that simplify, standardize, or routinize aspects of design, as when they size timber elements in a residence or select door hardware for an office.

The Role of Drawing in the Design Process

> It is through drawing that we not only explore the possibilities of new design but also acquire the fundamental language (the "meta-language") of architecture.
>
> *Simon Unwin (2014)*

> [R]elatively unstructured drawing or sketching is important in the early stages of the design process and ... more structured types of drawing occur later in the process.
>
> *Purcell and Gero (1998)*

Architects have a deep respect for drawings, and for the act of drawing, to the point where the physical act of drawing and the cognitive processes of design are often linked (Scheer 2014). Describing their own design processes, architects frequently speak of "talking" to a drawing, giving the act of drawing a conversational character and the drawing a role as dialogic partner. Ideas are explored by "drawing them up" using appropriate wall thicknesses, furnishing dimensions, etc. The drawing serves as an analytic platform, a means of overcoming human memory limitations, and as a means of experiential simulation. During schematic design, most drawings go through many revisions, though it is said that Frank Lloyd Wright drew the plans for his iconic Falling Water in the few hours it took the client to drive from Chicago to Taliesin, and that he talked to himself as he drew, describing how the clients would experience the house (Tafel 1979).

Drawings also have meanings beyond the explicit delineation of geometry— latent content. They capture, express, and test ideas, but within the practice they also demonstrate prowess, persuade, explain, and organize work (Robbins 1994). Drawing is part of the identity of architects, and has undergone radical transformation in the last two decades. Among the changes: Most drawings are done using computers; drawing sets contain more 3D details; senior management is less able to jump in and help with production during times of heavy work; and the infrastructure costs of supporting additional designers have grown due to required hardware, software, and training.

Drawings also have implicit meaning for contractors. In the past, they might look at a hand-drawn detail and estimate the architectural knowledge and experience of the designer who drew it based on the drawing's lettering and line quality. The assumption was that drawing skill develops along with professional experience, so a well-drawn detail is also a well-informed detail. While CAD drawings are more uniform than hand drawings, they do manifest their author's drawing skills and knowledge; it just takes time to learn to read them.

Finally, because clients often interpret hardline drawings as complete and unchangeable, designers may use software to add squiggle to the drawings, imitating back-of-the-envelope sketching and reminding clients that decisions are mutable.

Drawing Conventions Organize Cognitive Work

One of the significant, and unstudied, changes which digital drawing technology has brought about relates to the relationship between the designer and their visual field. Traditional drawings, on large pieces of paper, adhered to a preset scale, which establishes a visual field within which the drawing is built. Detail finer than a pencil point cannot be represented, and spatial organization larger than what can be viewed standing at the drawing board is difficult to think about.

A change in scale is both cognitive and graphic. You cannot put much detail on a 1:100 scale drawing, and a pencil-width line is probably all you need to represent wall thickness. A 5× or 10× scale change permits a change in representation detail, from single-line walls to double-line walls, to details of framing and trim. This change accompanies, or is accompanied by, a change in focus, from arrangements of space to wall locations to wall composition to assembly details.

The Role of Memory

> One of the most powerful aspects of human thought is our ability to sort objects and experiences into categories and then apply our knowledge about the category to deal with the case at hand.
>
> Earl Hunt (Hunt 2002)

Individual designers use memory to draw on personal experience in many ways. Current problems may well bring to mind solutions developed in either the recent or distant past and personally experienced or studied by the designer. Alternatively, models of use, society, and construction developed by others and encountered through study may be recalled at an opportune time, suggesting a solution to a problem. Finally, during projects designers acquire a great deal of client, site, and project information that may not be explicitly recorded, but which influences the design development.

Design firms, and their many individual designers, collaborate on projects through formal and informal memory systems. Preliminary sketches are shared. They become formal documents recording decisions, but throughout a project alternative solutions jostle to become (or be reinstated as) part of the current solution. Desktop "litter" speaks to passers-by. Past presentations and projects decorate the walls. Projects develop a narrative; memory of design rationale is important in preserving decision-making coherence over the course of a project, and, of course, memory is the foundation for learning.

Design by Analogy

We often rely on solution by analogy, finding an example in history or experience that is like the current problem and borrowing or adapting the previous solution. We might call this a *problem-space search*, in contrast to a solution-space search. Its prevalence in practice might account for the observation that architects, unlike scientists, often reach their full power later in life, once they've had time to acquire enough experience.

When you've solved a problem once, the "not obvious" solution of Chapter 6 becomes available as an object of memory, which means future presentations of the problem have obvious or easier solutions. In writing about the problem of problems, Kees Dorst describes five levels of expertise, corresponding to "five ways of perceiving, interpreting, structuring and solving problems." These are novice, beginner, competent, proficient, and expert, the last of which "responds to specific situation[s] intuitively, and performs the appropriate action, straightaway" (Dorst 2003). Similarly, in his book *Notes on the Synthesis of Form*, Christopher Alexander (1964) distinguishes between unself-conscious design, as practiced in cultures where tradition dictates how to solve problems, and self-conscious design, in which changing conditions are so challenging that reference to history is inadequate and the designer must actually posit a new solution.

Design as Selection

While inventive creativity is important, in many ways design is about selection—selection of strategies, materials, configurations, manufactured systems and objects such as door hardware, lighting systems, appliances, and carpet patterns. Building types such as "split-level ranch-house," "semi-detached house" or "three-story walk-up" convey a lot of information about the design, information drawn from a foundation of life experience and a pool of culturally validated examples. Without copying any particular design line for line, the designer may use such memory to organize a new response and create a new design. Memory and experience help guide selection, as does knowledge embedded in books such as *Building Construction Illustrated* (Ching 2014), *Architectural Graphic Standards* (AIA 2016), and the venerable Sweets® Catalogs (Sweets 2016), all of which provide condensed and simplified selection opportunities. In addition, most firms maintain detail libraries of tested solutions to particular configurations, and it is common for designers to review, or mine, past projects for ideas.

It is worth noting, in passing, that the increasing availability of digital fabrication technology has released (or challenged) designers to design where they might have selected in the past. This, in turn, has led to proliferation of unique buildings and objects and has both advanced and benefited from the use of sophisticated 3D modeling technology, illustrating the degree to which these topics are intertwined.

Digital support or fuel for a memory-based design process suggests one of several directions: *experiencing* or *presenting* architecture through digital media, *recording* personal travel in text, photos and sketches, as well as *cataloging* and *recalling* travel or design history.

Problem Solving Through Subdivision

When memory and the related strategies discussed above fail, as they often do in at least some aspect of design, designers fall back on various strategies that aim to recast recalcitrant problems in ways that are more tractable. Plato's suggestion to "separat[e] the Idea into parts, by dividing it at the joints, as nature directs"—what we might call *divide and conquer*—is one of the most common.

Modern attempts to formalize design processes are rooted in time-and-motion studies on mass-production assembly lines and the field of operations research, beginning in the early twentieth century, as well as more recent investigations of artificial intelligence. The time-and-motion approach tends to view the whole as simply the sum of the parts. Once isolated, each partial process could be streamlined on its own. Industrial production in World War II demonstrated the power of the divide-and-conquer approach. By dividing problems into smaller, more malleable, parts, solving them individually, and aggregating the results, it was possible to resolve many challenges, while taking advantage of collaborative multi-member teams of average people. Individual action and ingenuity took a back seat to methodological rigidity and the study of process.

What isn't always appreciated is that the very act of dividing the larger problem into smaller ones involves selecting one solution paradigm over other possible approaches. As Thomas Kuhn explored in his *Structure of Scientific Revolutions* (1962), it is possible to force-fit old solutions onto new problems right up to the point where the evidence of mis-fit becomes overwhelming and a paradigm shift occurs as a new model replaces the old.

Applied to routine problems, in the absence of overwhelming evidence demanding a paradigm shift, this approach often works adequately. It also describes the system of disciplines, consultants, and contracts that generally governs the process of design, and which is supported by most current software systems. These, in turn, correspond fairly well to the cognitive frameworks found in the IBIS and CBR approaches, as we will see shortly. There are, however, research directions that challenge this model, where the whole is more than the sum of its parts.

Developing Memories

Memories do not simply happen; they are built. The common admonition that young designers should "see the world" and sketch or take notes in their sketchbooks clearly seeks to augment their store of available memories as well as their sensitivity to cultural nuances through reflective study. Traditional paper sketchbooks permit travelers to construct a powerful blend of text, diagrams, and pictorial drawings at an almost infinite range of scales. Digital alternatives in the form of tablet or laptop computer, which preserve both sketch and text, might facilitate subsequent search or cataloging, but have thus-far proven elusive, though photo-sharing, and internet technology in general, certainly represent a giant step forward.

Personal travel, while valuable, is an expensive and time-consuming process. Historically, efforts have been made via words, painting, still photography, and film to convey experience of places and times not otherwise accessible. More recently, various attempts have been made to use internet-based technologies and sensory immersion to enable *cyber tourism* (Prideaux 2015). Such digital travel is not limited to the present—there are digital reconstructions of the ancient city of Pompeii as well as the Taj Mahal.

Many of these digital media recordings or reconstructions are unsatisfying. In addition to being limited to sight and maybe sound, they are often difficult to navigate without drifting through walls or floors and becoming disoriented because they lack kinesthetic feedback; in some cases they are nauseating. Head-mounted displays are notorious causes of motion sickness, but even flat-screen displays are susceptible when viewing very wide-angle images. There is a need for developers who understand how people visualize and move through the built environment to improve both the media and technology for these experiences. Numerous initiatives employing video-game engines such as Quake or Unity have made only slight progress, but rapid evolution of digital camera and display technology continues.

Nor is all this travel divided cleanly between real and virtual. The availability of location-aware interactive mobile devices such as smartphones and tablets has enabled the deployment of augmented reality systems that can "watch over your shoulder" as you travel and offer information about buildings, menus, shops, sights, history (e.g., http://etips.com). Some also allow you to post location-specific comments of your own, for subsequent travelers.

The recent explosion of personal photography has made image cataloging, tagging, and organization software widely available, and a few researchers have sought to leverage those collections to support design creativity (Nakakoji *et al.* 1999).

Collective Memories

Design organizations have collective memories built on the memories of individual designers present in the organization, but augmented by stories, project archives, and manuals of standard work, and supported by libraries of reference material drawn from the larger professional culture. Being outside the individual designer's head, many of these projects utilize computers to store and retrieve information, augmenting individual memory in complex settings.

Case-Based Reasoning

Individual personal memory is fallible, and when discussing design we often care as much about human motivations and decisions related to an environment as we care about the physical product. Capturing intent, in words, can be important. According to its proponents, *case-based reasoning* (CBR) "means using old experiences to understand and solve new problems" as a systematic activity (Kolodner 1992). It is a problem-solving technique that uses similarities between current problems and past problems and their solutions to propose, critique, and build solutions to current problems. Past experience (cases) are recalled, compared to current circumstances, adapted as needed, evaluated through feedback and/or simulation, and repaired (learned from) for next time. Computational challenges exist at each of these steps, but substantive CBR systems have been built.

Issue-Based Information Systems

A related approach focuses on the problems or issues that emerge during design, explicitly capturing their argumentation, decomposition, and resolution. The resulting *issue-based information systems* (IBIS) have been used to address very dense design environments with many interacting and possibly conflicting design issues, arguments, and resolutions (McCrickard 2012). By structuring knowledge of known problems and their extant solutions, the approach attempts to organize and apply experience to new problems in a systematic way. PHIDIAS, one such system, offered an interactive graphical environment for capture, review, and resolution of issues and design rationale connected to NASA spacecraft design (McCall *et al.* 1994).

Pattern Language

Memory, acting through personal association, or structured through CBR or IBIS, provides individuals and groups with tools for addressing complex design situations. On a more analytic or reflective track, designers and theoreticians have sought to organize or analyze human-built environments in order to identify the "atomic" design elements of configuration. Christopher Alexander and colleagues developed the theory of what they called a *pattern language* by segmenting, characterizing, and categorizing the oft-used parts, or patterns of design. A single pattern, #242, might be a "front door bench" (Alexander *et al.* 1977). A complete design consists of an assembly of elemental patterns, organized and orchestrated to work together. In the context of design computing it is interesting to note that computer science has embraced the concept of pattern languages in interface design as well as algorithm design and programming instruction.

Space Syntax

Attempting to understand larger organizations of human-created space, researchers at University College London and Georgia Institute of Technology, have developed *space syntax* theories and processes (Hillier and Hanson 1984). From the SpaceSyntax.net website:

> Space syntax is a science-based, human-focused approach that investigates relationships between spatial layout and a range of social, economic and environmental phenomena [including] patterns of movement, awareness and interaction; density, land use and land value; urban growth and societal differentiation; safety and crime distribution.
>
> *(Space Syntax 2015)*

Understanding that buildings are both meaningful from the outside as civic or cultural products, and meaningful from the inside to the inhabitants that work in and move through the spaces enclosed by the building, the concept of space syntax seeks to understand how the texture and organization of space in human habitation interacts with social and cultural factors in predictable and analytic ways (Scoppa and Peponis 2015).

Coupled with the development and application of computer-based geographic information systems (GIS) to major urban areas world-wide, historical research, and a theoretical orientation towards spatial metrics such as street width, sight lines, walking distance, and pedestrian behavior, space syntax researchers use large data sets to support their analyses, as well as observational data from video cameras and other sensors, to advise civic and private developers on many real-world projects (e.g., http://spacesyntax.com).

Shape Grammars

Individual designers and historical eras often develop stylistic consistencies in the visual or topological character of their buildings. Thinking of these buildings as cultural utterances in a language or design, it makes sense to take a linguistic approach to understanding architecture. George Stiny and James Gips first developed and publicized their *shape grammar* theory in 1971 (Stiny and Gips 1971). The approach seeks to identify, isolate, and characterize sets of *generative rules* for producing new members of a particular oeuvre (Stiny and Gips 1971; Stiny 1980). Each rule specifies an initial condition and a replacement condition. The process of applying rules begins by searching a design representation for conditions that match the initial condition, or left-hand side (LHS) of a rule. Where multiple rule matches are found, a choice must be made as to which rule to apply. Often this is left to a human operator. Once a choice is made, the system replaces the portion of the representation matching the LHS with the replacement condition, or right-hand side (RHS), after which the rule-matching starts again. Examples of architectural grammars that have been developed include Palladio's villas (Mitchell 1990), Queen Anne houses (Flemming 1987), and Frank Lloyd Wright's Prairie-School Houses (Koning and Eizenberg 1981). While the process may sound automatic or even arbitrary, the generative rules of the grammar can be seen as embodying significant design knowledge, and their production can be likened to design itself (Königseder and Shea 2014, citing Knight 1998). While early grammar rules were derived by experienced researchers and expressed by hand, recent research aims to use a space syntax approach to uncover a grammar in existing work (Lee *et al.* 2015).

The Roles of Certainty, Ambiguity, Emergence, and Flow

> In Google's world, the world we enter when we go online, there's little place for the fuzziness of contemplation. Ambiguity is not an opening for insight but a bug to be fixed.
>
> *Nicholas Carr (2008)*

Some designers work with broad gestures and very soft leads—even charcoal or watercolor—in the early stages of design, resulting in representations that are open to substantial amounts of interpretation. While antithetical to the precision of construction documents, the resulting ambiguity can also be seen as a resource, inviting collaborative interpretation and participation in the design process, engaging clients and coworkers alike.

In contrast, the unambiguous quality of crisp laser-printed drawings has often been cited as a problem when presenting preliminary design ideas to clients—who perceive the design as more immutable than it really is. As with any interpersonal exchange, designer and client communications need to be fluid and comfortable to tease out the full depth and extent of thought. Projecting too much certainty too early can cut off discussion, leaving reservations and concerns unuttered, possibly causing friction or expensive changes later.

The problem exists at a deeper level as well. As we will see in Chapter 8, external representations are also a designer's means of communicating with their "future self." Most designers approve of ambiguity early in design, but recognize that it must ultimately be resolved. This might tempt us to view the design process as simply the systematic elimination of ambiguity. Digital systems might seem to fit the task admirably since they thrive on specificity, on precise taxonomies that distinguish rather than blur. However, ambiguity seems to be an important component of emergence as well—"the process of converting an implicit property into an explicit one" (Talbott 2003, quoting Gero and Yan 1994). Emergence happens when a representation invites and is subjected to reinterpretation. In Figure 7.2 the left side represents "two overlapping rectangles." On the right, after only a slight reinterpretation, are the "two L-shaped polygons" implicit in the first shape, but moved slightly apart. While visually similar, their data representations are quite different, as we'll see in the next chapter. Further, this isn't the only possible emergent form, creating two problems. Can the software discover emergent alternatives to the current representation if asked, and might it possibly *prompt* reinterpretation by suggesting them? This turns out to be one of the sub-problems of shape grammars (interpretive flexibility is necessary in order to match grammar rules broadly). However, given the complexities of representation and the many possible reinterpretations, supporting reinterpretation and emergence is certainly going to be a challenging problem.

Parameters and constraints can interact in complex ways. William Mitchell (2009) observed: "building up parametric object families ... becomes a fundamental design activity, requiring deep architectural understanding, technical skill and considerable investment." Indeed, an architecture firm with a contract for a series of new library buildings tried to simplify their work by creating an extensive set of

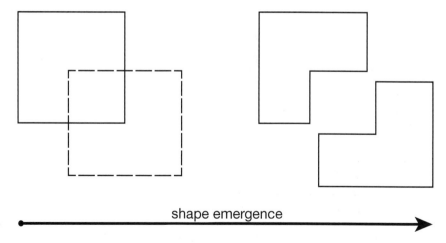

FIGURE 7.2 Emergence: A new interpretation of the representation requires new data.

parametric BIM elements during design of the first, only to find that the parametric constraints actually impeded their work on subsequent libraries and had to be removed.

The challenge of creating a family of related parametric forms is one of cognitive capacity; being able to think through the implications of multiple dynamic interactions. This is exactly the challenge of design. Complex user interfaces, no matter how conceptually elegant, can require so much cognitive overhead that there is little room left to pay attention to the design problem. Further, the attentive state of *flow* (Csikszentmihalyi 1991), in which a designer is able to fully engage a problem and often make significant progress, may well be as unattainable in a cognitively demanding environment as it is in a distracting one, including a complex interface or one that inhibits emergent reinterpretation. The joy of drawing, for many designers, resides in the power available through a very simple interface that requires reflection, but almost no other cognitive load. Reproducing this in a tool that also supports detailed building design at a semantic level remains one of the grand challenges.

Design as Social Action

Buildings and related environmental interventions are rarely an individual creation, but are, instead, a form of collective and collaborative human expression. They exist as cultural productions, or statements, in an evolving cultural and social context. A certain amount of ambiguity facilitates emergence of collaborative interaction, and an undisturbed interaction with the problem may allow useful reinterpretations to emerge. As a product of a shifting cultural foundation, the appropriate metrics for assessing the result may well be unknown in the beginning, and it is possible that their *meaning* in these contexts arises as much from the *process* through which they take form as from the particular dimensional and material choices of their production. Modern sociological theory sees *negotiated meaning* as a product of interaction.

Wicked Problems, Situated Action, and Puzzle Making

The particular solution and the problem definition fit together and emerge together from the process. This view of design as puzzle making (Archea 1987) or co-evolution (Dorst 2003) is very different from the more mechanistic divide, conquer, and aggregate view of design methods presented by Newell and Simon. On the flip-side, the *wicked problems* identified by Rittel and Webber can be seen to emerge from the mis-fit of problem and solution, arising from conflicts in the negotiation of meaning. Given the expense, permanence, and possibilities of most building projects, it is then unsurprising that at some level most design problems seem to be at least a little wicked.

Research into "situated action" by anthropologist Lucy Suchman (1987) and others (e.g., Heath and Luff 2000) has focused increasingly on the emergent character of action, arising less from individual cognition than from social interaction and technological affordance. While focused on the social and emergent aspects of action, this research does not entirely discount planning, which has a role in constructing the outline of the action. The ultimate relationship is likened to that of a jazz musician to the musical score.

Such research raises questions about the ability or long-term suitability of systems of pre-defined production rules to serve the emergent character of design, unless there is a mechanism for improvisation, though perhaps if the number of rules is large enough the designer can improvise within the system. The challenge for design computing is the amount of freedom needed. Pencil and paper, CAD, BIM, and a copy-machine exist on a continuum from complete freedom to complete pre-definition. At the deepest level we can ask *whether* systems can encode and capture "best practices" without limiting the designer to "standardized riffs." At a shallower level, we can develop strategies to best use existing constraint models and families of parts to provide productivity without predestination.

Digital support for puzzle making necessitates creating a system that does not require a problem definition to precede the pursuit of a solution, and which is able to move easily between editing one and editing the other, or perhaps makes no distinction between the two, but considers them two aspects of one production. Wicked problems are more difficult, because they challenge the notion that it is possible to generate or identify solutions through systematic action; they fall back on action mediated on communication.

Tools Interact with Cognition

Tool Neutrality and Cognitive Side Effects

> [A]rchitects' tools aren't neutral. They have particular affordances and biases, and they implicitly embody particular assumptions and values.
>
> *William J. Mitchell (Mitchell 2009)*

There is a persistent tendency to treat technologies as interchangeable, distinct from the human cognitive processes—something we might call a theory of *tool neutrality*. It is implicit in the argument often made in the 1990s that CAD was "just another pencil." Yes, 2D CAD was brought into most offices as a one-to-one replacement for hand-drafting and was thought to require largely the same knowledge and skills and produce the same results. However, studies of technology projects in other environments have demonstrated the complex interaction of digital technology with human cognition and social interaction patterns in many work settings (Lakoff and Johnson 1980).

The magnitude of the change wrought by the introduction of IT into architecture offices came home to me a decade ago when visiting a medium-sized local firm. Their monthly IT bill had just passed their monthly rent—in an industry where drafters used to bring their own tools and needed little more than a door-on-sawhorses drawing board. In an industry where status is frequently linked to drawing prowess and where senior office staff could "jump on the boards" and draft during a deadline push, CAD tools have redistributed power and introduced new paradigms to production, such as "pair work" (Hancock 2013). Whether these changes are simply transitional or persistent remains an open question.

Infinite Zoom Variation

Traditional drawings were produced at only a handful of different scale factors. While final plots may still honor these conventions, the need to fit small and large parts of projects onto a relatively small computer screen has made precise scale factors a thing of the past during editing. As a side-effect, it isn't uncommon for students to produce dramatically out-of-scale elements during early design as they learn to adjust their visual estimates to the drawing.

The ability to zoom in, combined with the persistence of data, offers the designer a seductive escape from dealing with big-picture problems first. When stymied by a problem at one scale, they can feel productive by zooming in to a smaller part of the project and working on it in the meantime. Where conventional scale-changes might enforce attention to large-scale issues early in the design, the CAD designer needs to develop an internal discipline about problem decomposition and priority.

The Decline of Hand Drawing: Side-Effects

Designers generally work from the outside in—e.g., establishing organizing principles on large-scale drawings before allocating uses and configuring individual spaces or picking door hardware. In the past this sequence was encouraged because a change-of-scale required manually creating a new drawing. The cost of creating that somewhat redundant representation encouraged the designer to make the changes big enough to matter. Since scale changes also enable new information to be seen or shown, creating the new drawing also provided an opportunity for the designer to recapitulate, and check, the design logic of the smaller-scale drawing. Such recapitulation is lost in the CAD world.

More fundamentally than we think, the *act* of drawing, not just the *production* of drawings, may be important to design thinking. James Cutler, a prominent Pacific Northwest architect, came to us a few years ago looking for help with a research idea. He has run a successful award-winning office for many years, but in recent years had observed that his interns are less able to execute requests requiring that they independently develop the initial sketch of a design idea. He had discussed his observation with senior management at a number of offices across the country and found anecdotal consensus on the observation. His own conclusion is that modern students are not sketching enough by hand.

In a culture with a deep belief in the efficacy of digital technology, this concern nonetheless finds support in several places. First there is fMRI evidence that it matters whether a child learns to hand-write text rather than type it. Something about the neural activation of forming the letters by hand facilitates development of writing skill (Richards *et al.* 2009). Second, recent studies at the university level, based on the educational psychology theory of *desirable difficulty* (Bjork 1994), have found that students taking longhand notes learn better than those taking notes on a laptop (Mueller and Oppenheimer 2014). Finally, within the realm of architecture and design, research has shown that "conventional modeling methods ... satisfy the requirements for intermittent divergence, while parametric modeling methods ... undermine them" (Talbott 2003).

These observations suggest that rather than being neutral one-for-one replacements for traditional tools, digital tools are different in important ways that affect the cognition of their users, and that the most appropriate tool for an expert may well be different from that which best suits a neophyte.

Summary

Unsurprisingly, human cognition is complex. Design cognition is especially complicated by the existence of several quixotic phenomena, including emergence, creativity, and flow. Design process and design knowledge both contribute to solving problems, capturing and supporting process, and building personal and institutional knowledge. Constructing software systems that learn from users, accrue knowledge, facilitate browsing of design rationale and alternative design options, is a substantial challenge. Creating systems that do this without interrupting flow with complex interfaces and cognitively demanding interactions will be even harder. Doing it in a way that supports life-long learning and development will require skills from design, psychology, sociology, and human–computer interaction.

Suggested Reading

Archea, John. 1987. Puzzle-making: What architects do when no one is looking, in *Computability of Design*. Edited by Y. Kalay, 37–52. Hoboken, NJ: John Wiley.

Cross, Nigel. 2011. *Design thinking: Understanding how designers think and work*. Oxford: Berg.

Lakoff, George, and Mark Johnson. 1980. *Metaphors we live by*. Chicago, IL: University of Chicago Press.

Lawson, Bryan. 1997. *How designers think: The design process demystified* (3rd edition). Oxford: Architectural Press.

Norman, Don. 1988. *The design of everyday things*. Cambridge, MA: MIT Press.

Suchman, Lucy. 1987. *Plans and situated actions: The problem of human–machine communication*. Cambridge: Cambridge University Press.

References

Akin, Ömer, Bharat Dave, and Shakunthala Pithavadian. 1987. *Problem structuring in architectural design*. Research Publications, Department of Architecture, Carnegie Mellon University.

Alexander, Christopher. 1964. *Notes on the synthesis of form*. Cambridge, MA: Harvard University Press.
Alexander, Christopher, S. Ishikawa, M. Silverstein, M. Jacobson, I. Fiksdahl-King, and S. Angel. 1977. *A pattern language*. New York: Oxford University Press.
American Institute of Architects. 2016. *Architectural graphic standards* (12th edition). Hoboken, NJ: John Wiley.
Archea, John. 1987. Puzzle-making: What architects do when no one is looking, in *Computability of Design*. Edited by Y. Kalay, 37–52. Hoboken, NJ: John Wiley.
Bjork, R. A. 1994. Memory and metamemory considerations in the training of human beings, in *Metacognition: Knowing about knowing*. Edited by J. Metcalfe and A. Shimatnura, 185–205. Cambridge, MA: MIT Press.
Bryce, Nessa Victoria. 2014. The aha! moment. *Scientific American Mind 25*: 36–41.
Carr, Nicholas. 2008. Is Google making us stupid? *The Atlantic* (July/August).
Ching, Francis D. K. 2014. *Building construction illustrated* (5th edition). Hoboken, NJ: John Wiley.
Close, Chuck, Kirk Varnedoe, and Deborah Wye. 2002. *Chuck Close*. New York, NY: The Museum of Modern Art.
Cross, Nigel. 2011. *Design thinking: Understanding how designers think and work*. Oxford: Berg.
Csikszentmihalyi, M. 1991. *Flow: the psychology of optimal experience*. New York: Harper Collins.
Dorst, Kees. 2003. The problem of design problems. *Expertise in Design: Proceedings of design thinking research symposium 6*. www.creativityandcognition.com/cc_conferences/cc03Design/papers/23DorstDTRS6.pdf.
Dorst, Kees. 2011. The core of "design thinking" and its application. *Design Studies* 32: 521–532.
Flemming, U. 1987. More than the sum of parts: The grammar of Queen Anne houses. *Environment and Planning B: Planning and Design* 14: 323–350.
Gero, J. S. 1990. Design prototypes: A knowledge representation schema for design. *AI Magazine* 11: 26–36.
Gero, John (ed.). 2011. *Design computing and cognition '10*. New York, NY: Springer Science and Business Media.
Gero, J. S. and M. Yan. 1994. Shape emergence by symbolic reasoning. *Environment and Planning B: Planning and Design* 21: 191–212.
Hancock, Lillian. 2013. Visualizing identity: Perspectives on the influences of digital representation in architectural practice and education. Unpublished Master's thesis, University of Washington.
Heath, C. and P. Luff. 2000. *Technology in action*. New York, NY: Cambridge University Press.
Hillier, Bill and J. Hanson. 1984. *The social logic of space*. New York, NY: Cambridge University Press.
Hunt, Earl. 2002. *Thoughts on thought*. Mahwah, NJ: Lawrence Erlbaum Press.
Kalay, Yehuda, L. Swerdloff, and B. Majkowski. 1990. Process and knowledge in design computation. *Journal of Architectural Education* 43: 47–53.
Knight, T.W. 1998. Designing a shape grammar: Problems of predictability, in *Artificial Intelligence in Design '98*. Dordrecht: Kluwer
Kolodner, Janet. 1992. An introduction to case-based reasoning. *Artificial Intelligence Review* 6: 3–34.
Königseder, Corinna, and Kristina Shea. 2014. The making of generative design grammars, in *Computational making workshop: Design computing and cognition 2014* (DCC'14).

Koning, H. and J. Eizenberg. 1981. The language of the prairie: Frank Lloyd Wright's prairie houses. *Environment and Planning B: Planning and Design* 8: 295–323.

Kounios, J. and Mark Beeman. 2009. The aha! moment: The cognitive neuroscience of insight. *Current Directions in Psychological Science* 18: 210–216.

Kuhn, Thomas. 1962. *The structure of scientific revolutions*. Chicago, IL: University of Chicago Press.

Lakoff, George, and Mark Johnson. 1980. *Metaphors we live by*. Chicago, IL: University of Chicago Press.

Lawson, Bryan. 1997. *How designers think: The design process demystified* (3rd edition). Oxford: Architectural Press.

Lee, Ju Hyun, Michael J. Ostwald, and Ning Gu. 2015. A syntactical and grammatical approach to architectural configuration, analysis and generation. *Architectural Science Review* 58: 189–204.

McCall, Ray, P. Bennett, and E. Johnson. 1994. An overview of the PHIDIAS II HyperCAD system, in *Reconnecting: ACADIA '94 (Proceedings of the 1994 conference of the association for computer aided design in architecture)*. Edited by A. Harfman, 63–74.

McCrickard, D. Scott. 2012. Making claims: Knowledge design, capture, and sharing in HCI, in *Synthesis lectures on human-centered informatics*. San Rafael, CA: Morgan & Claypool.

Mitchell, W. J. 1990. *The logic of architecture*. London: MIT Press.

Mitchell, W. J. 2009. Thinking in BIM, in *Architectural transformations via BIM*, 10–13. Tokyo: A+U Publishing.

Mueller, Pam A. and Daniel M. Oppenheimer. 2014. The pen is mightier than the keyboard: Advantages of longhand over laptop note taking. *Psychological Science* 25: 1159–1168.

Nakakoji, Kumiyo, Yasuhiro Yamamoto, and Masao Ohira. 1999. A framework that supports collective creativity in design using visual images, in *Creativity and cognition 1999*, 166–173. New York, NY: ACM Press.

Newell, Allen, J.C. Shaw, and H.A. Simon. 1958. Elements of a theory of human problem solving. *Psychological Review* 65: 153–166.

Norman, Don. 1988. *The design of everyday things*. Cambridge, MA: MIT Press.

Osborn, A.F. 1953. *Applied imagination: Principles and procedures of creative thinking*. New York, NY: Charles Scribner's Sons.

Prideaux, Bruce. 2015. Cyber-tourism: A new form of tourism experience. *Tourism Recreation Research* 30: 5–6.

Purcell, A. T. and J. S. Gero. 1998. Drawings and the design process. *Design Studies* 19: 389–430.

Richards, Todd, V. Berninger, P. Stock, L. Altemeier, P. Trivedi, and K. Maravilla. 2009. fMRI sequential-finger movement activation differentiating good and poor writers. *Journal of Clinical and Experimental Neuropsychology* 29: 1–17.

Rittel, Horst and Melvin M. Webber. 1973. Dilemmas in a general theory of planning. *Policy Sciences* 4: 155–169.

Robbins, Edward. 1994. *Why architects draw*. Cambridge, MA: MIT Press.

Scheer, David Ross. 2014. *The death of drawing: Architecture in the age of simulation*. New York, NY: Routledge.

Schön, Donald. 1984. *The reflective practitioner: How professionals think in action*. New York, NY: Basic Books.

Scoppa, M. D. and J. Peponis. 2015. Distributed attraction: The effects of street network connectivity upon the distribution of retail frontage in the City of Buenos Aires. *Environment and Planning B: Planning and Design* 42: 354–378.

Space Syntax. 2015. Space Syntax Network. www.spacesyntax.net.

Stiny, G. 1980. Introduction to shape and shape grammars. *Environment and Planning B: Planning and Design* 7: 343–351.

Stiny, G. and J. Gips. 1972. Shape grammars and the generative specification of painting and sculpture. *Information Processing '71*. Edited by C.V. Freiman, 1460–1465. Amsterdam: North-Holland Publishing Company.

Suchman, Lucy. 1987. *Plans and situated actions: The problem of human–machine communication.* Cambridge: Cambridge University Press, *Construction building materials directory.* http://sweets.construction.com

Tafel, Edgar. 1979. *Years with Frank Lloyd Wright: Apprentice to genius.* New York: Dover.

Talbott, Kyle. 2003. Divergent thinking in the construction of architectural models. *IJAC* 2: 263–286.

Unwin, Simon. 2014. *Analysing architecture* (4th edition). London: Routledge.

8
REPRESENTATION
Capturing Design

> What he showed in this experiment was that the kinds of problem designers were likely to identify and solve were to some extent dependent on the kinds of representations they used.
>
> Bryan Lawson (Lawson 1999, citing Eastman 1970)

> To a man with a hammer, everything looks like a nail.
>
> Mark Twain

Designers use their minds to define and solve complex problems containing many elements, more than their limited short-term human memory can easily retain. To augment their memory, they create external representations, objects such as notes, sketches, drawings, or physical models. While these external representations are often thought of as being the design, each captures only aspects of it. In this way the designer is able to grapple with complex configurations and pay attention to different elements of the design without losing track of information while their attention is elsewhere. Representations have additional value when they can be used as the foundation for computational evaluation or analysis of the design proposal, or as communication tools by which to convey design intent to clients, contractors, or other interested parties.

During the process of creating a representation, a semantic payload, or meaning, is attached to or embedded in the design artifact, a process that may be awkward and frustrating or fluid and effortless, depending on the fit between the designer's goals and skills and the characteristics of the medium. At a later time the designer can retrieve their ideas or concepts about the project, organizing principles, etc. by viewing the representation. Recorded meanings may be explicit (words or dimensional drawings) or associative (diagrams or words that evoke a particular memory).

Individual representations are also members of a *class* of similar artifacts, referred to with collective nouns such as *plans* or *sections*. As the epigraph indicates, the particular representation is not a neutral repository for information; the choice of representation and subsequent problem solving are linked.

Historically, drawings have been the dominant representational form in architecture. Sketches done quickly as part of early design exploration may indicate only important features of the site context, or basic relationships between project parts. Subsequent schematic drawings may give those relationships spatial form and approximate geometry, but these will be redrawn during design development with greater care for relations, scale, and spatial sizes. While each of these may be a "drawing," they serve differing cognitive roles, involve different levels of geometric precision, and draw on increasingly deeper reservoirs of project knowledge. It seems unlikely that a single tool would be suitable to all of them.

By augmenting our short-term memory, external representations allow us to engage complex problems, but each is created to address a particular set of issues, and is thus incomplete—just one of a (potential) set of related representations. It is not the only way a particular meaning might be conveyed, and it requires some skilled interpretation. It is a *rendering* of the designer's intent, created within the cultural context of the design office, expressing their style. It is incomplete in any of several ways, perhaps showing only part of the building, or focusing on the spatial layout but not the structure, etc.

Many alternative representations exist and are, in some sense, equivalent and interchangeable. This makes conversion between representations, and coordination of multiple representations, important tasks. Traditional design workflows locate the designer at the center of the resulting confederated representation, responsible for interpreting and checking information against the central design intent. This gives the designer great power over the final product, but requires significant time and expense and hinders implementation of digital processes. Since the vast majority of design work is now done digitally, it seems desirable to shift this responsibility by creating a single central digital representation (e.g., a BIM) from which others can be constructed or extracted as needed. Establishing such a representation and working out appropriate ways for all industry participants to create, interact with, and refine them is one of the core challenges of design computing, and faces a number of hurdles.

This chapter looks at representations in terms of their relationship to the cognitive task, underlying primitives, organizing relationships between parts, compactness, convertibility, expressive range or power, and level of abstraction. Good representations can be examined, or queried, to find out things we didn't know before. They may be shared with others, edited or extended, stored and recalled later. Good representations facilitate design by presenting appropriate abstractions and tools for different phases of design, making it easy to make changes at a level consistent with the design's evolution. A good representation communicates clearly to others, while inappropriate ones confuse discussion; but few, if any, representations are appropriate to all designs or all aspects of a design.

Representation and Cognition

Why do we use representations? The usual answer is that the limitations of human memory make it hard to keep track of all the details. Working memory is generally limited to 7–10 items of information. Recording information in written or drawn form preserves it. It also allows hierarchies containing related information "chunks" to be developed and manipulated. In this way, representations help us organize the cognitive task, linking our present and future selves.

Every representation is created by expressing an internal mental model in the vocabulary of the representational medium. A painter may imagine colors that don't exist, but can't paint with them. At the same time, knowledge of color theory and a supply of primary colors may enable them to create colors not yet available in the paint box. Awareness of the vocabulary and knowledge about how to stretch it to fit current needs are parts of the process. The painting emerges through an exchange of information between the painter's head, the canvas, and the paint box.

Of course, if the goal of design is Simon's "action aimed at changing existing situations into preferred ones," strapping on a tool-belt may seem more direct. It is, but it might not be the most efficient way to make change happen; where you can speak to a few people directly, an email expressing the same ideas can go to thousands. Where you can pick up a few boards, your mouse can indirectly move mountains. That's because representations can be used to communicate to others, leveraging our time and energy.

Efficient action is nice, but correct action is even better; remember the old advice to "measure twice, cut once." The design process uses representations to estimate the consequences of proposed changes before they happen, avoiding mistakes, cost, and injury. Representations must offer the means to preview, or simulate, consequences.

In the end, the rationale for expressing a building in any form except the finished product must identify *power over the design* as one of the main motivations. Designers create external representations to overcome limits of human cognition, in order to test and refine the proposal, and in order to clarify and communicate design intent. When changes are easily implemented and easily reversed—as when rearranging furniture in a room—we might simply make a change and see if we like the result. In such a situation, the design representation (the furniture layout in the room) is identical to the design problem. There is no abstraction involved; we can immediately evaluate the result visually, though we may have to guess about other aspects—such as how it will look in different seasons, or how well it will work during a party.

When it is not easy to implement or reverse a change, we often employ an external representation, such as a drawing or model, to help us design. As indicated in the epigraphs, the particular external representation chosen (whether digital or not) is important because it influences workflow, productivity, information management, and design cognition.

Representation and Analytical Power

As described in Chapter 2, computational tools combine a representation or *model* (information) with *algorithms* (processes) that change or act on the representation. Representations are static, while processes are dynamic. The two are tightly coupled, a bit like nouns and verbs in written language. As with language, effective human expression and communication requires control of both the representation and the algorithm, and they can range from the nuanced and elegant to the crude and obtuse.

Consider, for example, the way most of us were taught to compute sums on multi-digit numbers:

$$1024 + 516 = ?$$

The traditional algorithm for computing the sum goes something like this (given that we have memorized single-digit sums): start in the 1s place; add $4 + 6$ for 10. As 10 requires two digits and we only have room for one, record the 0 in the 1s place of the result and carry the 1 over into the 10s place. Next, $2 + 1 + 1$ (the carry-over) for 4 in the 10s place, with no carry-over. Now add $0 + 5 = 5$ in the 100s place. Only the first number has an explicit value in the 1000s place, but there is an implicit 0 in the second number, so add $1 + 0 = 1$ for the 1000s place. Final result: 1540.

Interestingly, this algorithm works equally well if the numbers are expressed in base two (binary numbers):

$$10000000000 + 01000000100 = 11000000100$$

However, it does not work well if the numbers are expressed as Roman numerals, where the order of characters in the representation is more important than their position:

$$MXXIV + DXVI = MDXL$$

The same quantitative meaning exists in each case, but the chosen representation influences what you can do with it and how easily; different representations require different algorithms.

A more graphical example arises in the context of subdividing large areas into smaller ones. In a landownership application you might consider each piece of property to be a polygon. This is the way the landowner tends to think of it—in terms of the boundary of their property. However, the different edges of a property boundary are *shared* with each of the neighbors. If each boundary is stored separately as a distinct polygon, problems may arise. For example, it is hard to guarantee that there aren't slivers of overlapping or unassigned property since the boundaries are duplicated and stored in different places. Further, if someone decides to give their neighbor a 10 foot (3 m) easement, the information needs to be stored in two places in the database, and updating it means finding the two boundary polygons. Fortunately, the *winged edge* data structure commonly used in geographic information systems is ideally suited to keeping track of boundaries such as this, and can be used to rapidly identify the unique boundary polygon of each property while being certain that all land is accounted for and there are no overlaps.

High- and Low-Level Primitives

Every digital representation is created from a vocabulary of fundamental parts, or *primitives*, each with certain qualities or *attributes*. These are designed into the software, represented via patterns of bits in memory. They are also what distinguish between BIM, CAD, and paint programs. These primitives are stored in a data structure that establishes simple or complex relationships among the parts, including lists, graphs, and hierarchies.

An application might use high-level primitives (e.g., "rectangle"), or it may employ low-level primitives (e.g., "line") to accomplish the same visual effect. User interfaces may mask the distinction by providing a tool that appears to draw a rectangle, but actually creates four lines in the drawing database. It may seem like a trivial distinction, but it dramatically changes what the user can do with the representation. The opposite sides of a rectangle remain parallel while a collection of lines need not, so a rectangle cannot be made into an arbitrary parallelogram. A rectangle is a closed shape while four lines need not be, so a rectangle can have a fill-pattern as an attribute, while the collection of lines may not.

Representation and Embedded Knowledge

The information in a digital representation, the data and its organization, is purposeful. There are specific ideas of how it can and will be used in a productive workflow. Every program is built around a particular representation, combining knowledge and process. The representation can encode disciplinary knowledge about the problem the software is meant to address, while the software's algorithms encode process knowledge.

Figure 8.1 illustrates this by comparing two representations that produce the same graphic figure—a rectangle. The left-hand representation embeds knowledge in the data by naming the primitive (RECT) and uses knowledge about rectangles to parameterize the figure using the coordinates of two diagonally opposite corners. The right-hand side shows an alternative representation constructed of four LINE primitives, each defined in terms of its endpoint coordinates. It is straightforward to extract and draw the four edges of a rectangle as lines, if needed, as shown in the lower left portion of the figure. The four lines, on the other hand, are not required by the rules of geometry to form a rectangle, but they might. The *knowledge* (RECT or not) missing from the representation can be created by using *process* to determine whether the lines satisfy the geometry conditions shown on the lower right. For this reason a rectangle is said to be a *higher-order* primitive than a line.

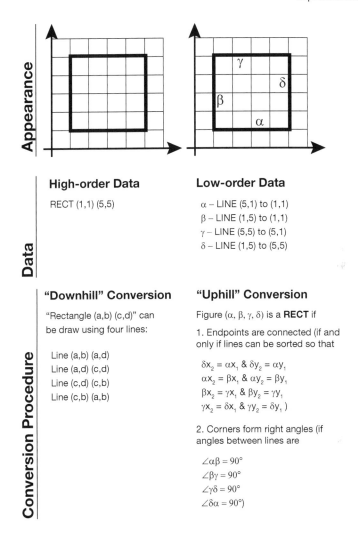

FIGURE 8.1 Alternative representations may look similar, but behave differently when edited.

The knowledge embedded in a system's representation can be seen as guiding the user to appropriate choices (Clayton 2014) and aiding productivity by reducing attention required to complete the task. It might also be seen as coercive or overly restrictive, depending on how well the representation fits with the designer's cognitive model, and how the user interface is implemented.

It seems that as representations become more focused and incorporate more specific disciplinary knowledge, their expressive range narrows, but they become more powerful and efficient within that disciplinary envelope. Simple low-level representations, such as those embedded in basic paint and draw programs, require greater cognitive involvement on the part of the designer, but offer a wider expressive range. If this observation holds, it may be that as we seek to embed more information in a unified building model, our interactions with it will become ever more ponderous and labored.

This may explain the simultaneous expansion of interest in BIM tools, whose highly developed representations may both empower and uncomfortably constrain design exploration, and scripting, in which the designer, using simpler manipulations of lower-level primitives, gains expressive freedom that might not otherwise be available. The cognitive engagement that accompanies such work also echoes the theory of *desirable difficulty* which has shown that students taking pen and paper notes in lecture classes reword content and develop deeper understandings than those with laptops, who tend to produce verbatim transcripts (Mueller and Oppenheimer 2014).

Representation, Affordances, and Task

Consider how your spreadsheet differs from your word processor. They both store text and numbers. You can position the insertion point in either one, and begin typing text. You can select sections of text and make them bold or italic. Both programs will add headers and footers to your pages. But your spreadsheet does not (happily) do word-wrap, so you must pay attention to line length and use the enter key to start a new line when needed. Changes during editing—insertion or deletion—will force you to re-flow the contents of your paragraph. Now, think about building a budget in your word processor. It will let you define tables, select cells and type category names and numbers, align columns, and even do simple math on the numbers, but it won't change the number of digits after the decimal point in a column of numbers, and it won't do complex calculations. The word processor and the spreadsheet share certain features, including elemental data types and simple formatting, but the relationships they enforce between the parts are different—they present different affordances to the user. They each offer good representations for their primary tasks, but rather weak representations for the other task, which is probably why both applications remain distinct tools on our desktops.

Now consider the relatively simple task of cropping a raster image. A high-end application such as Adobe Photoshop can do this easily, but a new user may find it difficult to locate the appropriate tool amid the many options available in the interface.

It is necessary to match task, representation, user interface, and cognitive state of the user in order to produce a positive experience. Doing so with a single tool that spans from the fuzziness of early design to the precision of detailed design is likely to be a challenge. The field of *user experience design* (UxD) has emerged, in part, to address this challenge.

Common Representations and their Problems

Chapter 5 described the state-of-the-art as regards commercial CAD and BIM systems, but if you ask a designer what they're doing with their notebook and pen or pencil/stylus, you're likely to get an answer with words like "concepts" and "sketching ideas" or "exploring relationships" or maybe "working out a problem." If pressed, they'll probably reference "forces" and "activities" or "organizing principles." These words describe the conceptual contents that their drawing or sketching media are used to capture, the design elements they work with cognitively while their pen or pencil makes marks on the page. One reason that pen and paper remain popular is that, while simple, they may be used to construct complex representations with very different semantic content. While individual computer programs may deliver the power of particular analog activities, from free-form doodles to 3D sketching and BIM, each imposes its own set of requirements and limitations on the designer. Delivering the fluidity of interaction and range of expression possible with pen and paper remains a substantial challenge.

The characteristics of the medium and the features of the representation artifact are intimately connected. As a consequence, another understanding of the term *representation* (or sometimes, *model*) has emerged. It is used in the collective sense to refer to categories of similar objects—plan drawings, pasteboard models, 2D CAD data, BIM data, water colors, etc. This is especially relevant to discussion of software, where the cost of development invariably requires that it support creation of multiple models in multiple projects. This goal creates a preference for data features or operational affordances that supply the most widely sought-after, generic qualities of the medium, and may lead to provision of affordances that generalize or constrict representations for the sake of efficient production.

Closure and Complexity

It may seem that the best strategy for making an all-purpose design tool is to include as many high-level primitives as possible. The ones the user needs will be available, and the ones they don't need will simply not participate. There are two problems with this: closure under editing operations and interaction between primitives.

For example, several years ago there was a 2D drafting program that included lines, circles, arcs, and ellipses among its primitives. It also had commands to subdivide, or *break*, lines. A broken line turns into two shorter lines. A broken circle turns into an arc. Unfortunately, it wasn't possible to break ellipses, as there was no elliptical arc primitive. The set of primitives was not closed when using the *break* operation.

While closure is not necessary for the software to run, it is highly desirable. As the number of possible operations and the number of primitives each increase, assuring closure under all operations is a problem of geometric proportions. That is, it grows as the product of the number of distinct primitives and global operations.

Similarly, the complexity of some routine operations, such as line-of-intersection computations in a 3D modeling application, depends on the number of possible primitive types involved. If all models consist of polygons, we need only consider the intersection of two polygons, but if we have both polygons and cylinders, we must have code for plane–plane, plane–cylinder, and cylinder–cylinder intersections.

What this means is that completely aside from issues of interface complexity and user training, software developers are strongly motivated to use simple generic primitives in their representations rather than complex domain-specific ones. One of the attractive features of NURB modeling, in addition to precise surface representation, is that NURBs can be used to represent both planar forms and curved ones, so a NURB–NURB intersection computation, while complex, can be used in almost all calculations.

Drawings

It is worth remembering that the early efforts to apply computing to *design* occurred prior to the development of high-performance graphics displays, and actually tended to focus more on issues of optimized circulation and rational spatial layout than on appearance or documentation (Lawson 1997). The focus on drawings and computer aided drafting (the modern sense of the term CAD) came about with the explosion of personal computers in the 1980s and 1990s.

While they dominate the discussion, drawings play such different cognitive roles at different stages in the development of a design that it might be better to develop a more refined taxonomy. A bubble diagram is not the same as a plan sketch, which is not the same as a detail; there is a context switch that identifies which information in each drawing is important, and in what ways. Still drawing can support the design process in almost every stage.

Pencil and paper drawings do have a number of limitations. Changes in scale require re-drawing. The drawing can run off the edge of the paper if overall size, scale, and orientation are not considered from the very beginning. Even simple edits usually require erasing at least part of the drawing and redoing it. The big problem—the drawing coordination problem—is that nothing inherent to the medium guarantees that the building shown in one view is the same size as the building shown in another, or has all the same features (doors, windows, corners). Nor is it certain that window openings are in the same place in plans and elevations, etc. This problem is only resolved through significant expenditure of cognitive effort. Finally, even when drawings are complete and well-coordinated, most analyses require substantial human interpretation of the drawing before they can be completed (Augenbroe 2003).

Technologically, tracing paper, xerography, Mylar drafting film, light-tables and pin drafting all served to reduce the problems associated with assembling coordinated drawing sets. In this light, the adoption of 2D CAD and BIM are part of a natural progression, as both attack a number of consistency and efficiency problems head-on.

As discussed in Chapter 5, digitally a "drawing" usually means a graphic defined as a list of individual elements (lines, circles, arcs, and text), each defined by certain coordinates and parameters, while a "painting" means a grid of pixels, each assigned a color. In three dimensions, polyhedral or parametric-patch boundary representation or solid models capture geometry, with surface qualities or attributes assigned as parameters, images, or parametric textures. Object-based geometry takes up less space than raster data, and can be selected, scaled, copied, rotated, and manipulated parametrically. This foundation was natural for 2D and 3D CAD, at least in part because the actual decision of what to draw was left up to the human user.

Symbolic Content in Drawings

Architectural drawings contain representation of both the building geometry and a variety of cultural symbols, a fact which complicates the processes of learning to read drawings as well as the automatic production of drawings which is necessary to bridge between a 3D design model and 2D construction documents. For example, walls in plan are shown as if cut approximately at waist height, but this may vary in order to expose significant features such as partial height partitions, clerestories, mezzanines, etc. Even when the positions of the lines forming the boundaries of the wall can be determined geometrically, the poché between the lines is a symbolic and conventionalized representation of the wall assembly. Figure 8.2 shows the traditional poché for a wood-stud wall.

Elsewhere on the plans, one may find "dollar signs"—*symbols* representing electrical switches in location but not size or geometry. They are shown outside but geometrically adjacent to the wall in which the implied electrical box housing the switch should be placed. The same switch, shown in an interior elevation, may well have the geometry of a cover-plate. Note how the location of the symbol is geometrically relevant to the design, but the graphic of the symbol varies by drawing type (plan, elevation, etc.). In other cases, it is even more complex: doors are quasi-geometric symbols. They show where the door fits in the wall, how big it is, the volume intersected by its swing, and where the hinge goes—all geometric qualities. However, by convention doors are shown open in plan but closed in elevation, with a visible (but non-physical) arc in plan to indicate the swing, and (usually) a text label linking the symbol to additional information about fire rating, finish materials, hardware, etc.

Dimensions and pochés tell us things about the design, but not the geometry. Line weight identifies cut geometry and relative distance to the next surface; line style may indicate edges above or below; and color/shade may well indicate elements to be removed as well as added. All-in-all, construction documents follow a complex language of denotation. The notion that they are simply slices through a 3D geometry is clearly incorrect. BIM software, which does produce 2D drawings from models, generally employs a "box model" that permits or requires alternative 2D representations to be created for those tricky objects and assemblies that don't follow geometric drawing rules. Users are insulated from this complexity by a robust supply of pre-made generic building parts.

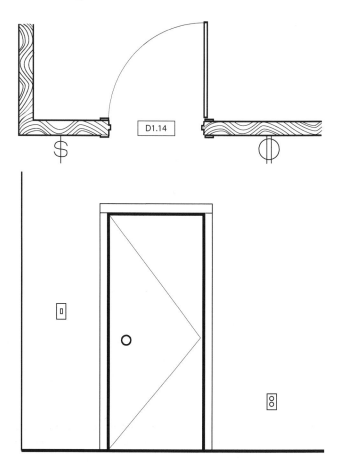

FIGURE 8.2 Symbolism and geometry in architectural drawing.

Buildings

The BIM packages of today are primarily design documentation systems, which means their primary product is construction documents. They represent buildings as needed to service that end. They include many pre-made building elements (wall assemblies, windows, doors, furniture, etc.) that have been created to work with each other and with the underlying drawing-extraction system. The predefined parts mean users don't have to address geometry vs. symbol problems just to get their first project done, but the complexity of these parts also means the users may not be readily able to delay decisions (e.g., walls must have a specified thickness), or add their own building components, making design exploration and experimentation more difficult.

Regardless of how drawing conventions expect doors to look, the contractor has to order and install real doors with real hardware and trim. Details and door schedules are intended to bring the relevant information together in a compact form, but they also duplicate information shown elsewhere in the drawing set, creating opportunity for errors. On a big project, just updating the door schedule can be a full-time job. In typical 2D CAD software, standardized symbols can help because they can be tracked and counted, but they can also be *exploded*—turned back into simple primitives during editing—which destroys their semantic value. Representing a *door* as an irreducible first-class object, with appropriate parameters for hardware, glazing, fire-rating, etc. creates data that can be reliably searched for, collected, and manipulated by software. It reduces errors and saves money. Combined with other data, it might enable code-compliance checking. On the other hand, there are many ways that doors may be constructed—pocket doors, sliding barn doors, double-doors, Dutch-doors, saloon doors—so making a universal "door" object in a BIM program isn't trivial, even if you know how to draw each of those.

Alternative Representations

The dominant characteristic of most vector representations is that of a list. Discrete data objects representing constructible parts of the building appear in no particular order within the list. Editing consists of adding, altering, or deleting elements, sometimes (via object snaps) after extracting coordinate data from existing objects for use in new ones. However, there are other ways of representing a design.

Topology: Beyond an Aggregation of Shapes

In some building types (especially hospitals, courthouses, and factories) there is a lot of attention to the proximity of activities, and the flow of people and resources. Issues of connection trump issues of location. Even in more mundane buildings, such as homes, adjacency diagrams showing relationships between spaces—bubble diagrams—play an important role in focusing client–architect communications on issues of acoustic and visual privacy, access, and spatial hierarchy. In these drawings, as in Figure 5.4, spaces are often represented as simple circles or rounded rectangles. Openings, hallways, and doorways are lines linking spaces. These are graphs. Landownership, floor plans, street networks, and many other basically 2D phenomena have been explored as graphs. Such representations make it possible to distinguish *inside* from *outside* and guarantee enclosure for energy calculations or 3D printing, and enable testing for code-compliant exit pathways in a high-rise, detecting unassigned square-footage in a rental plan, etc. They are not, however, widely used because they are hard to reliably extract from a large list of elements that may not join up exactly at their corners, etc.

The tendency to focus on a list of things that must be installed or constructed ignores the most important part of architecture—the spaces created between the walls. If the model is maintained as a list of objects, created in the order entered by the user and not organized according to the building geometry, it means the programs may not be able to answer questions involving spatial connectivity,

116 The Grand Challenges

circulation and emergency egress, or HVAC. Such information can be incorporated in a space network, adjacency graph, or "soap-bubble" diagram. Focused on spaces and their shared boundaries, such descriptions may also require different (i.e., non-manifold) topology for their full expression.

These models take advantage of connectivity. As illustrated in Figure 5.4b, if we establish the coordinate at each corner or intersection of a 2D plan, the walls connect pairs of points. If a shared point is moved, the connecting walls in all related rooms are adjusted. Doors are simply penetrable parts of the walls. We have added *topology* information to the model in the form of a *connectivity graph*. As shown in Figure 8.3, the graph can be used to establish room adjacency and possible circulation paths.

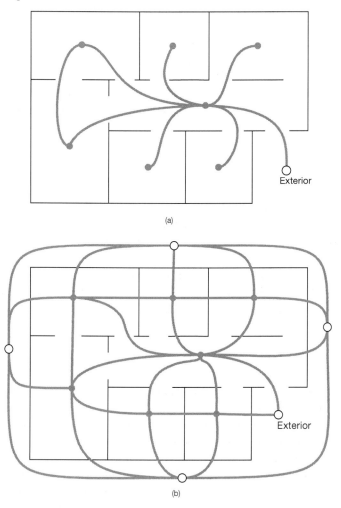

FIGURE 8.3 Planar graphs represent spatial relationships of (a) circulation, and (b) adjacency or boundary locations in a simple plan.

A *graph* is a concept from discrete mathematics. It is a set of nodes (the dots are points in space) connected by links (the walls). If you did a diagram of a website, the pages would be the nodes and the anchors linking between pages would be the links. A floor plan may also be thought of as a graph, as may a family tree. Graphs exhibit various interesting properties that are important to some aspects of design computing, such as circulation.

While the visual character of the building depends on representing the material that is there, circulation within the building depends on what is *not* there. The exact geometry of the walls (glazed or not, interior or exterior) isn't important. Within a space it is assumed that we can move about freely. Connections between spaces depend on where the doors, stairs, and elevators are. Doors connect two (and only two) spaces. This means that circulation can also be represented as a graph, with nodes representing spaces, and links (or edges) representing connections between spaces. If you overlay a circulation graph with its associated plan graph, they are clearly related. In fact, they are said to be *duals* of each other.

Graphs are useful for capturing something that is divided in irregular ways, such as the way land is divided by property-lines, and for representing circulation within cities or corporate campuses, where the nodes might be buildings or intersections and the connecting links are streets or sidewalks.

In a well-formed and complete model, this information can be computed from the boundary-element information for simple cases such as enclosed rooms connected by doorways, but it is much more difficult to extract in the case of environments with incomplete modeling or errors, or where there are implied spatial subdivisions such as pilasters, ceiling beams, flooring changes, or built-in furnishings. Extraction and interpretation of such information from the geometry of an environment is one goal of *space syntax* research, using techniques such as *isovists* (Benedikt 1979) and connectivity graphs.

In the 1970s there was a great deal of interest in the potential of linear graphs for space planning (Grason 1971), an interest that seems to crop up once a decade or so (Arvin and House 1999). Today, we see this technology in our route-finding satellite navigation systems.

Shape Grammars: History, Shape Equality, and Emergence

The multiplicity of equivalent representations possible in a line-oriented drawing editor makes it very difficult to establish equality between two drawings, or to identify changes. It also means that semantically different representations may appear to be quite similar. A modest reinterpretation of the lines may suggest a very different design. In the process of drawing, aspects of the project that might not have been associated with each other become visible, or a drawing may reveal a relationship between parts of the project that had been considered separately. This discovery of unrecognized information or possibility in the project is known as *emergence*. We have seen this as a component of design cognition in the previous chapter, but it also presents a representational challenge, because the emergent reading of the drawing may suggest a design move that the original data does not readily support. The history implicit in the data may need to be rewritten.

In the previous chapter Figure 7.2 showed a plan sketched initially as two overlapping squares, which subsequently suggested a courtyard and two L-shaped buildings, and the possibility of opening up the courtyard to the larger site. Most designers and most software would capture the initial sketch as two rectangles. In today's CAD software their overlap is not explicitly represented, nor are the points where they intersect. Reinterpreting the sketch as two facing L-shapes with a void between them requires the rectangles to be broken down into lines, their intersections found, and new polygon outlines created and substituted for the original squares. Since this is only one of several ways to reconstitute the outlines, there is no obvious automatic "rewrite" rule. It will probably be necessary for the designer to make a choice, although the emergent interpretation might be implied by the edit the user is attempting to make. These are open research questions.

The study of such transformations forms the heart of *shape grammars* (Stiny 1980), introduced in Chapter 7, an approach that evolved out of computational linguistics. While the most obvious characteristic of most grammar systems is sets of production rules from which increasingly complex configurations can be built, the underlying representational goal is one in which the sequence of construction doesn't matter, so history is removed, shape equivalence is computable, and emergence is directly supported.

In this context, one of the appealing qualities of a raster representation is its lack of history. In a paint program the squares of our example are just collections of pixels. While *we* may view them as squares, there is no drawing data that encodes their history. If the user can carefully select the pixels that constitute one of the two L-shapes, the job is done. The designer is free to reinterpret the sketch in any fashion at any time. Unfortunately, since this depends on the user, it also imposes skill and attention requirements on the user (to make the selection) and the use of raster data dramatically reduces the possibility of machine interpretation of the drawing. At this time, a representation capable of supporting the strengths of both vector and raster media remains elusive.

Decisions: Capturing Design Rationale

Representations focused on the constructed elements of a building do not capture design rationale. Designers and clients develop reasoning for decisions, but once a decision has been reached and recorded in the documents, there is no ready means of recording the reasons behind it—the office *debate* about where to put the CEO's office, or *why* certain areas have half-height partitions. These decisions are often made in response to specific client input or site conditions. In the face of complex and changing requirements, revisiting the reasons for a design being the way it is ends up relying on the notoriously fallible memories of the people involved, assuming they are still available to consult. This observation has motivated the development of *issue based information systems* (IBIS) (McCall *et al.* 1989) and *case-based reasoning* systems (Kolodner 1992), both of which feature representations that focus less on the physical aspects of a design and more on the problem structure and rationale that brought them to their current state.

Computational Design: Designing in Code

While geometry data can reflect thought, computer code can directly capture and express thought. Geometry editors such as 2D and 3D CAD/BIM systems take very neutral stances towards the geometry they edit; they are meant to be tools for the designer, not collaborators. Some formalisms, such as shape grammars, do capture elements of design thinking because they attempt to capture the rules (processes) that move a design from one state to the next. The appearance of scripting languages as a part of CAD software packages, beginning with AutoLISP in the 1980s but continuing to Generative Components, RhinoScript, Grasshopper, MAXScript, and Dynamo, taps into the persistent desire to explicitly create, edit, test, and execute design thoughts.

Modern scripting owes a debt to software systems that have been modularized and refactored to split user-interfaces from the geometry and rendering engines that operate in the background. In the process of making their systems easier to update and extend, developers have also made them easier for the user to manipulate, creating the world of scripting. Online open-source communities have disseminated excitement and knowledge about the resulting information ecologies, leading to a resurgence of interest in delineating design through code.

Parametric Design

One result of the interest in scripting in design is the rich and complex interaction of nested ideas and geometry that we call *parametric design*. These are objects, some of which might be quite complex, that change to fit the situation in which they are used. For example, the geometry of a real stairway is controlled by building code rules governing stair tread and riser dimensions as well as the overall floor-to-floor distance. Where earlier designers would have drawn a few standard-sized treads and computed the required number of risers from the floor-to-floor dimension and then drawn or noted the appropriate number of treads, today's designer can insert a parametric stair object that extracts the floor-to-floor distance from the system settings and draws itself fully formed, complete with code-compliant handrails. Further, if the floor-to-floor is large enough, it will insert a landing at the midpoint of the stair as well.

Mass Customization: Process as the Product of Design

The meaning, or semantic payload, of a representation is never entirely explicit. Much is implied, left to cultural norms of the construction trades (dimension round-offs, finish standards, etc.), or used and abandoned in the process of data entry (grid and ortho snaps, keyboard ortho constraints, relative coordinate keyboard entry, etc.). Further, and quite ironically, as digital fabrication becomes more common, the physical object is no longer the product of design, a data file is. In the even more abstract case of "mass customization" a family of objects, represented by end-user-specified parameters and designer-specified process, produces objects, some of which may not have been foreseen. The design itself becomes implicit.

Challenges to the Single-Model Concept

We know that the finished physical building will be singular, not plural, and believe that there is a single design toward which it is being guided. We know that efficiency can be found in computation processes that share data. These facts support the belief that a single representation is both possible and desirable. What challenges it? Certainly, industry inertia makes it hard to change, but are there, perhaps, more fundamental problems with the fundamental goal?

Incompatibility: Representation and Abstraction

Even within those representations focused on geometry, there is more than one way to represent a design project, meaning both that there is more than one form of external representation that might be used and that an individual designer's construction of an external representation will be personal and unique. Each representation is a means to an end, whether that end is design development, construction documents, rendering, energy analysis, cost estimating, construction planning, egress analysis, structural analysis, or all of these. As each of these ends has developed over time, they have developed their own representations for a building. Until quite recently, no one has sought to represent just a building, as such. That is what drawings were for, but drawings require interpretation, a step we are now seeking to eliminate. Arising in independent disciplinary silos, different analyses focus on different details of the building, ranging from surface color, to physical properties such as heat conductivity and strength, to properties of history such as embodied energy and carbon footprint, material sourcing, or operating expense. The analysis cannot be conducted if the information isn't available, but the information may not be explicitly stored in the data model. It may, in fact, be implied by the locale, the building type, building code requirements, or even basic industry practice.

Not only do the different analyses focus on different building details, different disciplines often use different approximations to produce analyses at each stage in the design process. And, while the geometrical representation of room surfaces required for construction begins fairly simply and gradually grows to be quite complex, some other representations remain fairly simple, even for completed buildings; for example, much HVAC design largely ignores internal partitions and treats the building as two zones—one near the windows and one interior. This geometry may not be explicitly defined by building elements, nor computable from those elements that are explicitly defined; requiring that it be supplied during analysis in order to complete the analysis. Such manual delineation means there will be multiple overlapping geometries in a building, that may have to be manually updated as well, adding opportunity for error. Nor are such updates trivial to make; when you move a wall it might matter to the economic model because it changes rentable area, but not to the HVAC system because it is interior to an HVAC zone.

Conversion Between Representations

While a single "cradle to grave" data model might be desirable, it is not yet available. Numerous circumstances arise in which designers need to convert data from one format to another. They may wish to use data supplied by the client, a consultant, or a vendor. They may wish to perform a numerical analysis or simulation of the project that their software does not directly support. Finally, they may find one tool conducive to schematic design, but wish to use another for design development. In each case, they will need to convert data from one representation to another. These conversions raise a number of issues.

Translation and Augmentation

The process of converting a design into an analysis model is likely to precipitate decisions. To create a cost estimate from a geometrical design representation we need to perform "material takeoffs" that turn CAD or BIM data into areas and volumes for estimating. It may also be necessary to explicitly define outlines for areas of flooring. To build a thermal-performance estimate, areas will again be needed, but carpeted floor may be treated differently from raw concrete (necessitating special/modified areas), most interior partitioning will be ignored, while exterior wall, floor, ceiling, and window assemblies must be specified thermally. Exterior walls will be aggregated based on assembly, orientation, and solar exposure, and both floor and ceiling conditions must be explicitly identified (occupied vs. unoccupied space? Above grade or below? etc.). While this information might be estimated or deduced from model geometry, other information, about occupancy schedule and occupant load (numbers of users and their activity level) and estimates of plug-loads from equipment, will almost certainly need to be added.

The point of these two examples is to illustrate how differently the building may be seen by different performance analyses. In each case the design must be interpreted in terms of the needs of the analysis, and in each case additional data must be supplied to the analysis. While these interpretations may be saved and reused in subsequent analyses, the limited integration of building data with analysis data is cumbersome.

Many Data Formats

Analyses and simulations may examine issues ranging from construction cost to structural soundness, to carbon-footprint, to energy use, to projected revenue generation. In each case computational models have been developed that incorporate the appropriate knowledge and process information. Each begins with the "building" geometry, but brings a specialized view of the design that often requires a different interpretation of that geometry and needs certain specialized information not found in the CAD or BIM model. The analysis thus requires both translation of the underlying design geometry and augmentation of that model with additional information. Each model begins with spatial information about the project (areas, lengths, volumes, locations, geometry, etc.) aggregating, simplifying and augmenting it to supply the necessary information. While supplying that data from the design representation would be sensible, and some vendors provide products that do this if you remain within their product family, it remains a very challenging goal in the general case.

One indication of the extent of fragmentation is the number of distinct file formats. One list of CAD and modeling file types includes over 1100 different file extensions for CAD data (file-extensions.org 2016). As mentioned in Chapter 5, efforts to establish a data exchange standard for the architecture, engineering, and construction (AEC) industry were not initially successful. While many programs are able to read and save the de facto standard DWG or DXF file formats, and are increasingly able to use Industry Foundation Class (IFC) or gbXML, the fact that these are not their native storage format means conversion is needed. While simulation platforms may import geometry from a CAD model, many utilize their own file formats for saving models and results. Thus, when a designer wishes to conduct one of these analyses, they must generally prepare input data for the analysis program, using their building design data for reference.

Phases and Focus

Individual design representations focus attention on those aspects of a problem which are open to manipulation during the design process, or which influence simulations or analyses that can be carried out based on the representation. In that sense, representations are incomplete. Further, as the designer's attention shifts, representations will shift as well, from simple to detailed, from occupant circulation to energy use, etc. In the past it was common to create new representations at key transitions or for specific purposes in the course of a design, as when sketch design drawings were "hard lined" on the drafting table, given typical dimensions, made orthogonal, etc. Or when material quantity takeoffs were done for cost estimating. Such transformations take time and energy. One of the persistent attractions of digital technology is the idea of representations that span a project from "cradle to grave," a feat that has yet to be fully realized. Among the heavy BIM users present at a 2008 GSA workshop in Seattle, most saw three broad categories of BIM model: design, construction, and operation—largely distinguished by the way in which contained information is used. Others speak of "federations" of BIM models (Scheer 2014).

Whether this is a single universal building model or a federation of BIM models, there will be a variety of conversion problems, either because the design must be re-created in a new representation or converted from one to another. This issue will re-emerge in regard to the role of expertise in the design process (Chapter 11), but consider this example from geometry: There are many times when a 3D modeling application needs to represent a surface with a hole in it, as when putting a window in a wall. As a wireframe elevation drawing in a CAD system, this can be done with two rectangles, one inside the other. If the outer rectangle has a poché or hatch pattern, the inner rectangle is usually expressed as a continuation of the perimeter of the outer rectangle, traversing one in a clockwise direction and the other in a counter-clockwise direction and drawing invisible lines on the edges that connect the two. This works okay in some surface-modeling programs too, in which a "hole" is just an area where no data is, but many hidden-line algorithms need convex polygons, or, worse, triangles. Converting our "wall with an opening" from one format to another may turn it into a wild collection of triangles. It'll render well, but if you edit it in this form and need to convert it back, it may be very difficult.

Design Iteration, Progression, and the Single Model (Round-Tripping and Incomplete)

As designs iterate to increasing levels of detail, their representations move from bubble-diagrams to rough wall layouts to detailed plans and sections. These changes illustrate that conversions happen over time within disciplines as well as across disciplines.

Each candidate design solution will consist of an arrangement of parts. Early in the process these might be rooms or zones. Later on they will be walls, doors, and windows. There are different possible representations of each solution. Each has its own appropriate level of detail for analysis or simulation.

Not all design media are used in a fully intentional way. With pen and paper, the designer's initial sketch may consist of unstructured, even random, marks on the page (Purcell and Gero 1998). These marks are thought to facilitate retrieval of content from the designer's long-term memory, content which begins to layer semantic structure onto the sketch, guiding its further development. Ambiguity and imprecise drawing thus engage long-term memory, possibly leading to emergence and a reassignment of meaning. This pattern of divergent thinking will be followed at some point by the designer shifting to a more analytical convergent mode of thought in which drawing tools and scaled lines play a more prominent role. At any time, the designer may revert to divergent ideational drawing, with the result that a given piece of paper may contain drawings of any type, all drawn with the same tools.

The digital tool-kit is currently less flexible, but arguably more powerful, divided roughly into the categories of sketching, drafting, and modeling, spanning from raster to 2D drawing to 3D modeling. Only modest long-term overlap occurs, though BIM and some 3D modeling systems do include fairly extensive drafting sub-systems, and "redlining" or similar markup tools may be applied to virtually any document that can be converted to PDF. The reduced flexibility comes from the necessity of picking a tool (program) at the beginning of a representation task and the limited range of expression that the tool facilitates. Greater power is associated with tools with a wider range of editing options (scaling, copy/paste, etc.), but few users sketch in a CAD program or draft in a paint program—transitions seem inevitable.

It may well be the case that the cognitive shift that occurs and the tool shift that is currently required as a design progresses happen at about the same stage in the development of the project, minimizing any negative impact attending to re-building the representation, possibly even facilitating development by encouraging a shift. Unfortunately, the iterative nature of design and the common occurrence of backtracking in some, if not all, areas of a project can make these transitions painful and frustrating. A system that blends the various representations, allowing quick overlay sketches in the midst of a BIM modeling session, has yet to emerge.

Conservation of Information

The concept of higher- and lower-level representations has been discussed previously. It is easy to convert a water color to pixels, but hard to convert one to drawing primitives. It is easy to turn a solid model into a boundary-representation model, and to turn that b-rep model into wireframe data, but it is much harder to construct a b-rep model from multi-view sketches, much less construct a compatible solid model.

The Round-Tripping Problem

This is particularly problematic when you consider that designers often "backtrack" to earlier design states as they work, with the result that re-creation or re-conversion may need to happen repeatedly. Further, the designer may wish to combine elements of a later state with those of an earlier one, which means that conversions may need to be bi-directional.

Traditional "dumb" representations such as paper drawings benefit from human interpretation that is intelligent, flexible, and adaptive. This enables a wide range of "downstream" uses of design data, but the interpretation step makes such use expensive. Digital representations, using computable semantically laden primitives in sophisticated relational data structures, promise to automate or eliminate interpretation and make downstream use inexpensive, potentially allowing design software to provide real-time estimates of cost, building use, carbon footprint, and energy use, all updated by background computations run against the single building representation in the BIM data set.

Unfortunately, we can eliminate interpretation only if the primary representation contains all information required by a secondary representation, but this is unlikely if not impossible. Automating interpretation will require various levels of information insertion, ranging from default values to full-on artificial intelligence. Automating interpretation may also require geometrical interpretation, projecting volumes from footprints of adjacent buildings, filling in detail in ambiguous areas, simplifying complex forms, or completing incomplete ones.

Summary

We create representations to exercise power over complex situations we need to change. Exactly how we conceive of them, as physical objects, relationships, or processes, and the assumptions we use to cage and control them matters a great deal. As participants in a collaborative social process, sharing our representations is increasingly necessary, requiring conversion and interpretation, simplification, and extrapolation along the way. The digital representations we create to work with them balance human hand–eye skills and associative memory against machine recordkeeping and process automation in processes where meaning changes as new interpretations emerge, as projects mature, or in response to a need to view the project in a new way. The ideal representation would support sketching, drawing, and constructing in a workflow where reinterpretation and emergence are supported, nagging is minimized, and designers are able to flip between thinking of spaces and surfaces, between organizing principles, regulating lines, and replacement rules or constraints, and where history can be ignored to follow an emergent idea, or consulted to recover design rationale. It is a task as complex as our personal and cultural relationship to the built environment as a whole.

No single representation exists at this time that can deliver all this, but Bryan Lawson closed out the third edition of *How Designers Think* with this relevant observation, which suggests how we might get there: "I have found that one of the most penetrative inquiries you make into how designers think is to demand that they use a computer tool and then allow them to complain about it" (Lawson 1997).

Suggested Reading

Lawson, Bryan. 1997. *How designers think: The design process demystified* (3rd edition). Oxford: Architectural Press.
Mueller, Pam A. and Daniel M. Oppenheimer. 2014. The pen is mightier than the keyboard: Advantages of longhand over laptop note taking. *Psychological Science* 25: 1159–1168.
Scheer, David Ross. 2014. *The death of drawing: Architecture in the age of simulation.* New York, NY: Routledge.
Stiny, G. 1980. Introduction to shape and shape grammars. *Environment and Planning B: Planning and Design* 7: 343–351.

References

Arvin, Scott and D. H. House. 1999. Modeling architectural design objectives in physically based space planning, in *Media and design process: Proceedings of ACADIA '99*. Edited by O. Ataman and J. Bermudez, 212–225. ACADIA.

Augenbroe, Godfried. 2003. Developments in interoperability, in *Advanced building simulation*. Edited by A. Malkawi and G. Augenbroe, 189–216. New York, NY: Spon.

Benedikt, M. L. 1979. To take hold of space: Isovists and isovist fields. *Environment and Planning B* 6: 47–65.

Clayton, Mark. 2014. Modeling architectural meaning, in *Building information modeling: BIM in current and future practice*. Edited by K. Kensek and D. Noble, 29–41. Hoboken, NJ: Wiley.

Eastman, C. 1970. On the analysis of the intuitive design process, in *Emerging methods in environmental design and planning*. Edited by G. Moore, 21–37. Cambridge: MIT Press.

File-extensions.org. 2016. 3D graphics, CAD-CAM-CAE file extensions list. www.file-extensions.orgfiletype/extension/name/3d-graphics-cad-cam-files

Grason, John. 1971. An approach to computerized space planning using graph theory. *Proceedings of the 8th annual design automation workshop*.

Kolodner, Janet. 1992. An introduction to case-based reasoning. *Artificial Intelligence Review* 6: 3–34.

Lawson, Bryan. 1997. *How designers think: The design process demystified* (3rd edition). Oxford: Architectural Press.

Lawson, Bryan. 1999. "Fake" and "real" creativity using computer aided design: Some lessons from Herman Hertzberger, in *Proceedings of Creativity and Cognition '99*. ACM.

McCall, Raymond, Gerhard Fischer, and Anders Morch. 1989. Supporting reflection-in-action in the Janus design environment, in *Proceedings of CAAD Futures 1989*, 247–259. Cambridge, MA: MIT Press.

Mueller, Pam A. and Daniel M. Oppenheimer. 2014. The pen is mightier than the keyboard: Advantages of longhand over laptop note taking. *Psychological Science* 25: 1159–1168.

Purcell, A. T. and John Gero. 1998. Drawings and the design process. *Design Studies* 19: 389–430.

Scheer, David Ross. 2014. *The death of drawing: Architecture in the age of simulation*. New York, NY: Routledge.

Stiny, G. 1980. Introduction to shape and shape grammars. *Environment and Planning B: Planning and Design* 7: 343–351.

9
INTERFACE
Where the Action Is

Where the Action Is.

Paul Dourish (2001)

As hard as it may be for modern computer users to believe in this era of smartphones, tablets, and touchscreens, early computers were not interactive. They were used to read files of input data, execute well-defined computations on that data, print or record results in new files, and then quit. A few examples of users actually interacting with a computer appeared during the 1960s as part of such projects as Ivan Sutherland's *Sketchpad*, but the event most often credited with launching the interactive-user paradigm is Douglas Engelbart's demonstration at the 1968 Fall Joint Computer Conference of work done at the Stanford Research Institute (SRI). The demonstration is often referred to as "the mother of all demos" in recognition of the number and range of innovations it showcased, including a variety of interactive applications, a mouse, a chorded keyboard, hypertext links, windows, and video conferencing (Engelbart and English 1969). The resulting paradigm shift paved the way for the Apple Macintosh, Microsoft Windows, and the WYSIWYG (what you see is what you get) paradigm that prevails today, in which the user continuously interacts with the machine to create or edit data, launch analyses, or view results.

From tools to accomplish specific narrowly defined computational tasks, computers have morphed into assistants with whom we communicate almost continuously about a wide range of daily tasks. This has been paired with sophisticated software and ever-greater use of computer-controlled devices, from electronic communications to robotic assembly and computer-numerically-controlled (CNC) fabrication devices. In this transformation, the "Run" command has been replaced by a complex interactive relationship with multiple virtual tools that demand, consume, and direct attention, mediate interpersonal relationships, and permit us to edit or control complex data sets and processes anywhere on the planet with the result that computers form an integral part of systems for air-traffic-control, manufacturing, subway systems, and buildings. We are increasingly aware of the importance, limitations, and affordances of the input and output mechanisms that constitute our interface with the machine, leading Paul Dourish, a prominent researcher from Xerox's famed Palo Alto Research Center (PARC), to title his book on the subject *Where the Action Is* (2001).

As computers become more central to design processes, and as buildings are increasingly recognized as significant computer input and output devices, the relevance of these topics to architects and other designers becomes more and more clear. This chapter will review some of the fundamental concepts, challenges, and opportunities of current and emerging interfaces linking designers to data, examining the direct manipulation paradigm, limits and use of gesture and vision, intent and meaning in drawing, computer support for design teams, and the potential of buildings as interfaces to their own operations.

The Interactive Paradigm

While Engelbart's chorded keyboard, on which patterns of simultaneous keystrokes permit efficient typing with one hand, never caught on, most modern desktop computers do use a mouse and keyboard for input, and create visualizations on one or more high-resolution raster graphics displays. High-speed networks link machines together, and many devices now include cameras and microphones for human-to-human communication. Mobile devices add more sensors, such as accelerometers and global positioning system (GPS) antennas, in a pocket-sized form-factor with high-speed wireless and cellphone network connections that allow these devices to track and monitor as well as interact with us. But let's focus on the interactive paradigm first. We will look at computers as communication platforms later in the chapter.

An interactive computer maintains some sort of representation that users manipulate ("view" or "edit") by communicating commands and parameters using mouse and keyboard, using what is called *direct manipulation* (Shneiderman 1982). Results are viewed on the screen and available for additional manipulation. The representation might be used as input for a separate computational process (e.g., rendering), or simply displayed, stored, and retrieved (e.g., word processing). In recent years this definition has expanded with the addition of voice and gesture input and "displays" (outputs controlled by the computer) that include building ventilation systems and smartphone icons, as well as traditional desktop graphics.

The issues associated with human–computer interaction (HCI) impact design in two ways: Designers interact with computers during design, as we have discussed in earlier chapters, and building occupants interact with surroundings that are increasingly incorporating sensors and computer systems that respond to their presence, as we will explore in Chapter 13. How meaning (commands or results) is conveyed between human and computer elements of design systems is of critical importance.

Design computing is concerned with both types of interaction under the broad subject of HCI, a subdiscipline of computer and information science that overlaps design computing in areas related to interaction and interface. While much HCI research is focused on efficient user interfaces for programs, researchers are beginning to address questions such as: Can your assisted living apartment help assess your cognitive state or turn off the stove if you leave it on? If each light in your house can be individually computer-controlled, and your house has occupancy sensors in every room, do you still need light switches? If you do away with switches, is occupancy enough, or should users talk to the lights, snap their fingers, or clap? Can a smart house learn from your past behavior and deduce what you need, perhaps holding phone calls or email if you are resting, or calling for help if you fall?

Deciphering Intent: Small Screens, Shaky Hands, and Fat Fingers

During design we use the computer interface to manipulate a representation of the design artifact, whether that is a drawing or a BIM model. The intended meaning, or semantic payload, of the representation is never entirely explicit. Meaning is often left to the cultural norms of the ultimate data users, or used and abandoned in the process of data entry. Further, as digital fabrication becomes more common, ironically, the object is no longer the product of design, a data file is. In the even more abstract case of "mass customization," a family of objects, represented by user-specified parameters and designer specified process, produces objects that may not have been foreseen. The design itself becomes implicit.

In the transition from hand-drawn to CAD-produced construction documents, one critic complained about the loss of "latent content" in the drawing of a detail. The claim was that a manually well-drawn detail with good line-weight control, lettering, and layout revealed a skilled hand, which implied a depth of experience and knowledge that the contractor could rely on. Because some physical attributes of a CAD-drawn detail (line-weight and lettering) no longer reflect the experience-level of the drafter, this critic felt they were inferior—the technology was artificially delivering something previously left to skill. However, while the observed change is real and CAD software does produce greater uniformity in drawings, skill and experience still reveal themselves in the organization of the drawing, the content shown, and in the operator's control of parametric features such as pochés or dimensions. In both cases there is more information conveyed than the face-value content of the drawing in question.

Designers organize space. Drawings as visual objects are largely about space, so drawings are an important part of design, but as we've seen in earlier chapters, neither CAD drawings nor BIM models are terribly good at directly representing space. Space is what happens *between* data elements. Interestingly, the user interfaces of most drawing and modeling programs don't directly support the organization of space either, though some emerging research is re-asserting the importance of such approaches (Jabi 2016).

Most drawing editors operate using *direct manipulation*, a term coined to describe the style of computer interface in which a user is able to select and manipulate elements of the display using only a mouse rather than typed commands (Shneiderman 1982). Most of us are familiar with direct manipulation in the form of word processor or drawing programs that highlight or display "handles" on objects as we manipulate them, even if we don't know the technical term. Direct manipulation makes the human user into the central change-agent in the data, able to observe each change both before and during the transformation. It also creates a precision problem.

The precision of direct manipulation depends on how well the editing gestures can be tracked, both by the computer and by the user. The geometry of the human body and limited muscle coordination mean most of us cannot draw a perfectly straight line—that's why there are straightedges. Nor can we draw easily two lines that end at precisely the same point. This is even harder when the screen you are viewing represents something the size of a building, and you require absolute precision. On top of this screen displays are pixelated, with each pixel representing a *range* of coordinates. Even if you have the hand–eye coordination to position the mouse cursor on the single pixel at the very end of an existing line, your point may not correspond to the same real-world coordinate as the existing endpoint. Because the lines mean more than the drawing, the *intended* result is often that the line needs to be constrained in some fashion above and beyond the direct sensing ability of the hardware.

Constraints reflect a designer's deeper understanding and organizational intent. But while designers may speak of symmetry, regulating lines and alignment in a design (Kolarevic 1994), drawing and geometry modeling environments control geometry largely using individual points in space, following the direct-manipulation metaphor. The enforcement of organizing concepts is left to the designer, not the software.

A limited suite of "pointing aids" in most CAD software facilitates precise pointing, including "ortho" mode, "grid snap," and the many "object snaps" (endpoint, intersection, etc.). These software filters (shown in Figure 1.3) correct for imprecise mouse, finger, or stylus input, constraining results and inserting meaning where hand–eye coordination fails. More advanced filters, or *inference engines*, allow software to offer hints or suggestions consistent with traditional organizing principles, displaying horizontal or vertical lines or endpoints related to points in the cursor's neighborhood or elsewhere in the drawing.

These input aids are often transitory. That is, in many programs "snapping" the end of one line to the end of another makes their endpoints coincident, but it doesn't create or encode the intent that they stay so. Similarly, making a line parallel or perpendicular to another, or offset a specific distance, doesn't mean it will remain so in the future. Such drawing aids can be used to efficiently make new data from old but without adding structure to the new data. This limitation is implicit in the representation of a drawing as a list of primitive elements. Fortunately, there is increasing recognition that constraints are important and explicit constraint manipulation is now available in some tools, though visualization and editing of constrained systems remains a work in progress.

Affordances and Seduction

The theory of affordances (Gibson 1986) leads software developers to make editing options visible within the interface. As a result, modern user-interfaces are festooned with tool icons and command menus. In the absence of a fixed design process, the interface offers a very broad range of tools to the user, based on the software designer's model of what the user is likely to do and the tools they will need. Well-designed tools incorporate disciplinary knowledge and culture in order to increase efficiency, but as they get more specific they also tend to be less flexible, and an inflexible tool cannot be used in unusual ways. The extent to which CAD or BIM software is guiding or limiting designer actions can also be a problem because users tend to do what is offered and may mistake action for forward progress in the design. Unlike the blank piece of paper demanding focus on unresolved problems, the software interface can invite action without focus.

Meaning in drawings arises from both conventions of architectural drawing (e.g., plans and sections are drawn as if cut horizontally and vertically parallel to primary elements of the building) and the specific contents of the image (a wall in plan is shown with a line on each side, at least in part because it is a vertical element and the plan-view line represents it well over its entire height)—a tilted wall would look quite different in plan. To the extent that a program seeks to enhance drawing productivity by assuming certain geometric relationships, or incorporating template graphics such as doors, toilet partitions, and cabinets that rely on convention, they subtly nudge designers in the direction of routine design. When using software such as BIM, that supports and enforces more complete architectural semantics and comes with many pre-defined parts including, perhaps, preset constraints, these benefits can only be realized if the design intent conforms to the expectations of those who created the constraint system, or if deviation is very easy to perform.

Even when an evolving design is very conventional, the designer may not have precise answers to wall-thickness or material-selection options at the time they sketch the plan. Tools that demand such answers before work can be done are frustrating, even when the answers are known, and can reduce productivity because they break concentration (Csikszentmihalyi 1991). Inappropriate nagging is one of the common complaints about BIM software today, contributing to the view that BIM is best applied to later stages of design, not during schematic design.

Where the Action Is

The field of design presents both one of the most difficult and promising opportunities to combine human and computer strengths. The complex mix of spatial details and cultural symbolism, of predictable pattern and delightful novelty, benefits from both a machine's ability to organize and retain detail and a human's integrative pattern making. The iterative process of design, operating across a variety of scales over a period of time, means that there is no tidy one-time handoff of responsibility from one partner to the other, or one representation to another. Instead, we see a complex interaction, a process of hypothesis, exploration, analysis, backtracking, and synthesis in which human and machine accept, process, and respond to information provided by the other. If the interchange can be organized well, it promises to be a powerful, fruitful collaboration that helps the designer to achieve the state of focused concentration and productive output called *flow* (Csikszentmihalyi 1991). Alternatively, designers could find their thoughts interrupted by ill-timed requests for information or offers of aid, a catastrophe illustrated by Microsoft's 1990s experience with an automated assistant in Word— *Clippy* (Robinson 2015; Gentilviso 2010).

Representations appropriate to the cognitive state of the designer are one challenge, but it is also important to uncover and make available modes of interaction that fit well with the work being done and others who may need that data. Over the lifespan of a building project, multiple participants will need to be brought onto the team and their input will need to be integrated into the overall project. Managers, bankers, clients, code checkers, and maintenance crews each have their own view of the building. Some users will be infrequent, or concerned with a defect or operation at a specific location in the building. Others may be monitoring progress during construction, or curious about the current state of the various building systems. Most of these won't find a BIM editor useful, but the information they consume and create might well be part of the BIM. Providing them with appropriate interfaces is an ongoing challenge.

During design, architects often straddle the boundary between digital and physical, frequently shifting back and forth between different representational media in order to gain insights and test conclusions about their designs. Sometimes these inquiries can be met by transformations within the medium, as when digital models are rendered, but sometimes they cannot. At such times it isn't uncommon to build physical scale models or prototype full-sized components. At a small scale, digital fabrication offers an enticing option—3D printing a physical model directly from the digital one. While far from instantaneous, inexpensive, or problem-free, this workflow is increasingly robust. The resulting physical model can provide tactile feedback, can be appreciated by multiple participants in a meeting, shared without technological paraphernalia, and modified with traditional tools if necessary.

However, it is also desirable to move from the physical world to the digital one, to (re)integrate design executed or modified in the physical world into the digital for further work; this is harder. Experiments with scanning, or smart construction kits that communicate their physical configuration to a computer (Gorbet and Orth 1997), or "over-the-shoulder" tools that build digital models in parallel with th e physical one (Hsiao and Johnson 2011) illustrate strategies for maintaining parallel models, while augmented reality and projection technology suggests useful strategies for including analysis results within real-world models (Ullmer and Ishii 1997; Hsiao and Johnson 2011), but there is more work to be done.

Even when a designer chooses to work exclusively with physical models, it may be desirable to execute simulation or analysis work on that model and view the results in the physical context (Ullmer and Ishii 1997).

In the future, when an architect is designing a house, their computer might use a camera to watch and interpret their hand gestures to indicate the overall shape, or let them consult "face to face" with a specialist half-way round the world, presenting both with a mutually controllable shared view of the design within which details of concern can be identified and discussed via video chat. As technologies develop, the designer might be able to give a client an immersive virtual-reality model of the design, but how do you represent different levels of commitment within the design? How do you distinguish between the "possible" and the "intended"? Immersive technologies—the hardware and software that enable them—are becoming commercially available, but the logic and cultural changes required to best fit them into both life and practice remains an area of research.

New Technology, Emergent Opportunities

As we attempt to extend the utility of design systems beyond the editing of drawings to manipulation of designs, certain challenges and opportunities appear. Some have to do with representation of non-geometric qualities in designs, such as the distinction between unambiguous aspects of the design (property line locations, topography, site features, utility connections, etc.) and more tentative or propositional content, or emergent reinterpretations of prior work. Visual representation of behaviors, such as parametric constraints, is another realm where progress is needed. Can the unambiguous be visually distinguished from the ambiguous, the certain from the uncertain, or the related from the unrelated? More research and development needs to be done in these areas.

Gesture and Voice

Another challenge has to do with memory, not as a design resource, but as a record. How can a design team remember the rationale for every decision as the design progresses, especially in the face of personnel changes? Now that computers can watch what we do and hear what we say, it seems only natural to include these modalities of interaction in our work with them. Can the rationales behind a design be easily captured and represented in digital form, to become part of the external representation, and searched or computed on in some fashion? Capturing time-stamped graphic and verbal components of design development in a recoverable and (more importantly) searchable form, without the tedium of typing descriptions or transcripts on a keyboard, might aid significantly with CBR, IBIS, and overall knowledge capture and review during a project (Gross *et al.* 2001; McCall *et al.* 1994).

Virtual Reality

Historically, architects have used renderings, models, and words to conjure up their design prior to construction. Now, virtual environment technologies enable a synthetic visual experience to be wrapped around the user, simulating the visual experience of being there. It seems only natural to adopt a technology that will allow others to preview design proposals, but the value goes beyond that. In an experiment conducted by an architecture student at the University of Washington in 1992, a virtual reality (VR) system was used to review models constructed in a traditional 3D modeling interface. The student found that the VR environment often revealed modeling and design errors simply because they were able to experience a wider range of views while using it (Campbell 1992, conversation with the author). As digital models become increasingly important in construction, what they "mean" becomes more important than what they "show."

Beyond simply presenting design geometry during design review, VR offers the possibility that individuals or groups might construct or edit models from within the virtual space. Perhaps we can even realize the common fantasy and edit the design by pushing walls around and drawing windows with our fingers! It's an interesting vision, and VR-based BIM will almost certainly appear, but there are a few problematic aspects that need to be sorted out—more design computing opportunities.

The Scale Problem

Virtual environments usually present the world at the same size as the user, but certain editing, such as shaping the terrain of a large site or laying out a large building, involves actions much larger than the human body. Virtual laser-pointers can be used to extend reach, but accuracy diminishes with distance, and without a "top" view it is hard to judge whether objects align, edges are parallel, etc. One possible solution is a "model-in-model" approach, in which a scaled-down version of the virtual world is located within that world, available to the user to view and edit (HITL 1996).

The Movement Problem

In first-person video games the game-controller provides relative movements to the protagonist's avatar or in-game representation—move forward, turn left, jump. To turn all the way around to our left we need only turn left enough times, all-the-while sitting in the same physical position in the world. This is fortunate, as the display screen that shows what our character sees remains solidly in one place as well. In immersive VR, on the other hand, viewpoint motion is usually extracted from the user's actual movements in the real world. Problems arise when the participant's virtual universe is larger than their physical one, causing them to bump into walls, or the virtual space includes stairs that don't exist in the real world. Various strategies to virtualize motion have been applied, ranging from placing participants inside gimbaled "bucky-balls" that rotate to accommodate "virtual movement" in any direction, to placing them in the seats of stationary bicycles or golf-carts whose controls update the in-world viewpoint (Mine 1995).

Disembodied-Participant Problems

The technology of VR necessarily involves obscuring the participant's view of the real world. This means they can see neither their own hands nor the facial expressions of other participants, whether physically co-located or not. While systems allow multiple participants to share a space and see each other in the virtual world, using avatars, the range of facial expression is limited, as is the range of motion that the avatar is able to sense and reproduce from their host user. This makes it very difficult to read a client or colleague's subtle emotional cues.

Augmented Reality

Similar to VR technology, augmented reality (AR) interfaces begin with a camera's live video stream and then digitally superimpose information over the top, as is commonly done with the "line of scrimmage" and "ten-yard line" in television broadcasts of American football. The video signal is monitored for specific colors. If they occur in certain recognized patterns, a secondary, usually synthetic, image stream is superimposed over the top of the live stream. This allows a graphic "marker" printed on a piece of plain paper to be replaced with correctly oriented and positioned real-time renderings of all or part of a design. Other marker-based tools can provide localized transparency or magnification of the model, access to layer visibility controls, overlay of analysis results, etc. (Belcher and Johnson 2008).

While possibly very useful during design presentations, this technology is also appealing to those who need data in the world, and is currently used in certain high-tech fields such as jet engine maintenance, or emergency response. It is also easy to imagine it in use during construction or remodeling projects in complex environments such as hospitals. Such applications are currently hindered by the difficulty of establishing position accurately and flexibly within a building where GPS signals are generally unable to penetrate.

Ubiquitous Computing

While interface issues are important to efficient and productive design processes, one reason we construct buildings is to support other human activities, including education, business, and housing. All of these activities are increasingly mediated by computing tools, tools which look less and less like free-standing desktop computers and more like elements of the overall environment. Researchers at the famous Xerox Palo Alto Research Center (PARC) were among the first to see this trend and coined the term "ubiquitous computing" in recognition of the shift from conscious (often expensive) computer use to casual background service (Weiser 1991). The continued decline in the price of computing, the increasing ubiquity of network connectivity, and commodification of advanced sensors like cameras, flat-panel displays, and touchscreens offer designers a new palette of tools through which to embed interfaces between people and tasks. With regard to interfaces between people and tasks, or interfaces between people and people, all this tech allows such interfaces to be embedded into the fabric of the building—into the architecture itself.

One Remove: Human–Computer–Computer–Human Interaction

The challenges of the interface are not limited to human-to-computer interaction. Computer networks allow people to interact, but they often limit those interactions in important ways. Gone are the casual encounters at the elevator, the overheard conversation that reminds you of a shared discussion topic, or the document seen in passing on someone's desk (Johnson 2001). Working in a group can be invigorating or frustrating, depending on how the members of the group are able to exchange ideas, interact to solve problems, and collaborate on shared tasks. Research in computer mediated communication (CMC) and computer supported collaborative work (CSCW) offers architects both a set of tools with which to practice and an area of research that might benefit from architectural metaphors and insights into human interaction and work.

Collaboration and Critique

Nor must the design computing researcher leave the office to find other opportunities to improve on the designer–computer interface. Traditional design presentations, with multiple drawing sheets pinned to a wall and a panel of reviewers, operate in parallel. Each viewer is able to peruse the entire presentation in any order, assembling understanding and questions from the information presented. If available, a 3D model may be examined from all sides—maybe even held in the hands. In contrast, projected digital presentations, even with high-resolution projectors, are usually built on the *slideshow* model. Serial and controlled from a single point, they come nowhere near their predecessors in terms of the amount of simultaneously available data, nor do they permit self-directed browsing of the information. If there is a 3D model, its presentation is either automated (as an animation) or requires use of the appropriate viewing software, a notoriously difficult proposition. Designers need new and better tools for this kind of interaction, tools that can be shared more easily than the single mouse or keyboard and which are simple and intuitive to use (Fritz *et al.* 2009).

Summary

Modern-day CAD and BIM work involves both human and computer. The efficiency or accuracy of analysis algorithms and the sophistication of representations only matter if the human and the computer can communicate effectively. HCI issues touch on fundamental cognitive issues of both individual designers and groups, as well as designer and client interactions with the design representation. As design data shifts from drawings produced for construction to BIMs produced to guide construction and operation, and as the digital design medium spans ever more of the building lifecycle, the amount of information in digital form and the number of stakeholders that need access to the data increases, while the suitability of existing authoring/editing interfaces to provide it for all users declines. We need richer models of what a BIM can be, and how it can be fluidly accessed and manipulated by these many stakeholders.

At the same time, the opportunity to embed sensors, displays, and actuators into the built environment is changing, and promises to continue to change, the form of buildings and the affordances they offer their occupants—areas of concern to architects.

Suggested Reading

Dourish, Paul. 2001. *Where the action is: The foundations of embodied interaction*. Cambridge, MA: MIT Press.
Shneiderman, Ben. 1982. The future of interactive systems and the emergence of direct manipulation. *Behaviour & Information Technology* 1: 237–256.
Ullmer, B. and H. Ishii. 1997. The metaDESK: Models and prototypes for tangible user interfaces, in *Proceedings of UIST '97, the 10th Annual ACM symposium on user interface software and technology*, 223–232. New York, NY: ACM.
Weiser, Mark. 1991. The computer for the 21st century. *Scientific American Special Issue on Communications, Computers and Networks*, 265 (3): 94–104.

References

Belcher, Daniel and B. Johnson. 2008. ArchitectureView: An augmented reality interface for viewing 3D building information models, in *Proceedings of eCAADe 2008*, 561–567.
Csikszentmihalyi, M. 1991. *Flow: The psychology of optimal experience*. New York: Harper Collins.
Dourish, Paul. 2001. *Where the action is: The foundations of embodied interaction*. Cambridge, MA: MIT Press.
Engelbart, Douglas and William English. 1969. A research center for augmenting human intellect. *AFIPS Fall Joint Computer Conference 33*, 395–410.
Fritz, Randolph, Chih-Pin Hsiao, and Brian R. Johnson. 2009. Gizmo & WiiView: Tangible user interfaces enabling architectural presentations, in *Proceedings of ACADIA 2009*, 278–280.
Gentilviso, Chris. 2010. The 50 worst inventions: Clippy. *Time*, May 27, 2010.
Gibson, James J. 1986. The theory of affordances, in *The ecological approach to visual perception*, 127–146. Hillsdale, NJ: Lawrence Erlbaum Associates.
Gorbet, M.G. and M. Orth. 1997. Triangles: Design of a physical/digital construction kit. *Proceedings of DIS '97: The 2nd conference on designing interactive systems—processes, practices, methods, and techniques*, 125–128. New York, NY: ACM.

Gross, M., E. Do, and B. Johnson. 2001. The design amanuensis: An instrument for multimodal design capture and playback. *Computer aided architectural design futures 2001*, 1–13. Dordrecht: Kluwer Academic Publishers.

HITL. 1996. The Greenspace project, Phase 2. Human Interface Technology Lab. www.hitl.washington.edu/projects/greenspace/

Hsiao, Chih-Pin and Brian R. Johnson. 2011. Combined digital & physical modeling with vision-based tangible user interfaces: Opportunities and challenges. *Computer aided architectural design futures 2011*, 785–799.

Jabi, Wassim. 2016. Linking design and simulation using non-manifold topology. *Architectural Science Review*. DOI: 10.1080/00038628.2015.1117959.

Johnson, Brian. 2001. Unfocused interaction in distributed workgroups: Establishing group presence in a web-based environment, in *Computer aided architectural design futures 2001*, 401–414. Dordrecht: Kluwer Academic Publishers.

Kolarevic, Branko. 1994. Lines, relations, drawings and design, in *Proceedings of ACADIA-94*. Edited by A. Harfmann and M. Fraser, 51–62. ACADIA.

Meyer, Robinson. 2015. Even early focus groups hated Clippy. *The Atlantic Online*, June 23.

Mine, Mark. 1995. Virtual environment interaction techniques. UNC Chapel Hill CS Dept.

Shneiderman, Ben. 1982. The future of interactive systems and the emergence of direct manipulation. *Behaviour & Information Technology* 1: 237–256.

Ullmer, B. and H. Ishii. 1997. The metaDESK: Models and prototypes for tangible user interfaces, in *Proceedings of UIST '97, the 10th Annual ACM symposium on user interface software and technology*, 223–232. New York, NY: ACM.

Weiser, Mark. 1991. The computer for the 21st century. *Scientific American Special Issue on Communications, Computers and Networks*, 265 (3): 94–104.

10
PRACTICE
Data, Documents, and Power

Architects have historically had great respect for and a very deep attachment to the process of drawing by hand, reflected in the many skeptical and concerned responses to office computerization offered by architects and educators such as Michael Graves (2012) and David Scheer (2014). In spite of these concerns, digital document production in the form of 2D CAD or 3D BIM has almost completely displaced hand-drawn documents, transforming the image of architecture and many processes of construction in the last two decades. The reason for this can be seen, in part, in the graphic shown in Figure 10.1, which is often attributed to HOK's chairman and CEO Patrick MacLeamy, though a similar graphic can be found in several mid-1970s texts, including William Mitchell's 1977 *Computer Aided Architectural Design* (Davis 2013).

The graphic illustrates the idea that an initially malleable schematic design becomes increasingly inflexible and expensive to modify as more and more interlocking decisions are made (and recorded in the design documents). Traditional document production techniques required significant manpower during the construction documents phase. Much of this effort went to coordinating redundant representations such as plans drawn at different scales, showing the same features in plans and elevations, and propagating changes through those various alternative representations—all relatively expensive processes. The promise of computing, especially BIM, is that a single data repository will be internally consistent and software can be used to perform the coordination required to produce the different-but-related documents, allowing more effort to be focused on design, thereby producing a better design at a similar or reduced cost.

This chapter will consider some of the ways in which practice is changing under the influence of digital technologies, changes that go far beyond the disappearance of drafting tables in offices. Digital tools have been introduced to practices in pursuit of greater productivity, in response to rising client expectations, and in

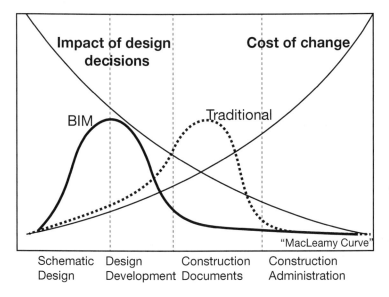

FIGURE 10.1 Motivating the shift to BIM: the "MacLeamy Curve".

recognition of new opportunities. Design computing technology presents the opportunity to respond to more data during design, to apply more computation to design decisions, and to explore morphologies whose complexity presents fabrication or construction problems. At the same time, opportunity does not guarantee results. Culture and practice changes may depend on the emergence of additional tools.

The Challenge of Doing Ethical Work

It is no longer enough to satisfy client budgets or aesthetics alone, not when as much as 40 percent of global energy is used in the production and operation of buildings whose construction may consume irreplaceable resources. Designers are being held morally responsible for community well-being and the architecture, engineering, and construction (AEC) industry is increasingly being challenged to demonstrate an ethical response to these facts through careful design and documentation, responsible material selection and reuse, appropriate detailing (e.g., for disassembly), and similar strategies. Designers are exploring ways of preventing or responding to natural disasters, even designing buildings that actively redistribute forces in the case of earthquake, tsunami, or flood, or that establish ad-hoc networks to supplement damaged information infrastructure. These concerns have placed additional demands on expertise, and increased the moral-hazard related to errors, whether that is building collapse, long-term exposure to carcinogenic materials, destruction of far-off hardwood forests, or simply inefficient cooling.

The Productivity Challenge

While the MacLeamy Curve is largely a rhetorical construct, an analysis of labor productivity data over the last 50 years by Stanford University showed declines in AEC productivity compared to US manufacturing overall (Teicholz et al. 2001). Though challenged on the grounds that it does not take into account a shift from on-site to off-site construction (Eastman and Sacks 2008), or reflect the challenges of an increasingly complex and demanding product (Davis 2013), a recent reconsideration reaffirms the basic claim that productivity in the AEC industry has declined, while general manufacturing productivity has increased dramatically (Teicholz 2013), largely through the application of automation and information. The implication is that the AEC industry is not making the most effective use of these technologies.

Large clients such as the General Services Administration (GSA 2007) and the Construction Users Round Table (CURT 2010) in the United States have responded to this conclusion by pressuring the profession to go beyond 2D CAD and make greater use of BIM in the belief that this will improve results. Architects, schools, and new students have invested in developing the required knowledge and software expertise. At the same time the traditional design–bid–build model of construction, in which designers and builders are often cast in antagonistic roles, has been evolving to make better use of digital technology through new practice models emphasizing collaboration. The iconic architect's drawing board and parallel rule have disappeared from most offices, completing the changes begun when 2D CAD was introduced. They have been replaced, at a much higher cost, by multiple large monitors connected to fast PCs licensed to run expensive CAD and/or BIM software.

While CAD and BIM have revolutionized the way documents are prepared, the role that paper drawings (or their digital equivalent, PDF files) retain as the dominant legal document in the building process is just beginning to change. The GSA now requires delivery of BIM data allowing automated checking of a schematic design against the design program. The emergence of new practice models such as integrated project delivery (IPD), which seeks to implement the MacLeamy shift by bringing construction expertise into the design process earlier, has begun to erode the traditional boundaries between design and construction. The result is that detail design and even fabrication now increasingly occur directly from design documents provided by the architect, eliminating a layer of shop drawings and data re-construction.

Thus, while it is hard to see how ICT could, in itself, cause a decline in productivity, the evidence certainly invites investigation into the ways design and construction knowledge are represented, prepared, compared, refined, and applied in practice. In the fragmented US AEC marketplace, the uneven productivity boost may well benefit some players more than others. Clearly, as architects are responsible for design coordination, the increasing complexity of buildings has made coordination both more critical and more difficult. Compounding this, greater emphasis on *performative architecture* (Kolarevic 2005), the effort to forecast and control project costs and performance, has placed additional documentation expectations on designers. Where digital fabrication makes unique design responses possible, integrated digital fabrication workflows may well have greater financial benefit to the builder. Adjusting compensation schemes will take time.

Within offices the increasing emphasis on the quantifiable aspects of design may also be causing a drift in the role of architects (Scheer 2014). Licensed for their attention to life-safety issues, but usually selected for their form-making expertise as well, architects have only recently embraced technologies of production as indicators of professional prowess. Prior to the arrival of CAD in offices, hiring a new employee did not require significant capital, as drafting tools were relatively cheap. Further, when project schedules required, senior members of a firm could contribute directly to drawing productivity by "jumping on the boards." Skill in drafting and freehand drawing established status within the profession, and skillfully drawn details were often ascribed higher reliability (Robbins 1997). Now that the production of legal documents has largely shifted to digital tools, there is stratification within architecture offices between those who can and those who cannot perform the work, as well as the emergence of new behaviors, such as *pair-work*, in which an entry-level technologically savvy user works alongside an experienced but technically less-savvy designer, thereby leveraging both digital skills and disciplinary experience (Hancock 2013).

Construction Process Changes

Today's building *process* is also changing. Where earlier surveying techniques required a time-consuming chain of positions linking a site to benchmarks, the *Total Station* electronic theodolites in use today can establish position and measure distances and angles with great precision, often under automated control. Such speed and accuracy, coupled with data drawn from the design database, have contractors talking about large-scale site work being conducted automatically using earth-moving equipment without human drivers. Digital technology enables contractors to fabricate and position building elements with great precision. Using these technologies, M.A. Mortenson, the contractor on the Denver Art Museum's Frederic C. Hamilton Building, designed by Daniel Libeskind, actually completed steel erection three months ahead of schedule and saved the city US$400,000 in the process (Mortenson 2005). The ubiquitous tower-crane operator follows a "lift schedule" that is carefully planned and optimized to deliver construction material to the appropriate part of the project at the ideal time, functioning very much like a very large 3D printer.

Easy access to correct 3D models by designers and contractors is more than a cosmetic benefit. Consider the challenge of correctly placing steel in formwork where it will be embedded in the walls of a concrete building core, possibly weeks before erection of exterior walls or placement of floor beams. At that stage in construction there is little else on the site to provide guidance, and drawings can be somewhat abstract. Use of 3D renderings during daily work planning has been found to reduce placement and orientation errors for steel connectors embedded in poured-in-place concrete construction (Campbell 2004).

Design documents are necessarily somewhat approximate or vague. Traditional 2D drawings may not explicitly represent the true shape of curved or angled building elements. In addition, contract document drawing sets bring together the work of many specialized disciplines, work that is often done independently and presented on separate sheets within a set. Finally, in order to allow for sub-contractor choice and contract alternatives, design documents may include some intentional ambiguity. While architects are usually responsible for coordination of the overall work, the final site of coordination is the building itself, and it is not unusual for construction projects to encounter spatial conflicts or clashes between systems such as reinforcing steel and embedded conduit in a concrete column, HVAC ducts and beams or elevator shafts, etc. Resolving such clashes in the field can be very expensive and time-consuming. Complete 3D models (BIM or otherwise) greatly facilitate clash detection. Further, some industry segments, such as HVAC contractors, can convert 3D data directly into fabrication data for automated tools. For this reason, contractors have often been keen to adopt BIM technology across the industry. Architects, in their coordination role, would seem to be the party that should be responsible for collecting or providing this data, but contractors may see the greatest benefit. Motivating efficient work and data sharing beyond simple contract documents needs to occur. Profit and risk sharing through non-traditional contracting practices might be one way to address this evolving situation.

Information Value: Sources and Sinks

When contemplating the productivity puzzle in light of the above, it is perhaps not surprising that the most visible group of early adopters of BIM and related technology has been contractors. Tools and data that reduce the work that must be done on-site pay dividends, and 3D digital models offer many opportunities to benefit. Many contractors will build their own BIM model as a means of testing their construction process, looking for clashes, establishing construction schedules, and subsequently managing sub-contractors, billing, and progress. The 3D model can be used in bending reinforcing steel, cutting structural steel, and fabricating sheet-metal elements for HVAC systems. In some more recent cases, contractors have relied on data to enable them to fabricate wall panels or entire units of the completed building off-site in climate-controlled environments, with greater productivity, reduced risk to workers, and improved quality.

Implicit in the pursuit of productivity gains is the reuse of data—often someone else's data. In a fragmented industry such as the US AEC industry, this requires trust and amounts to a risk and reward transfer, as noted above, often with the architect coordinating or providing information that others benefit from. Sorting out acceptable contracts and ground rules remains a work in progress, one that design computing might well influence or be influenced by.

Being able to access and use the design team's BIM for their planning and testing would be a great benefit to the contractor, but does not happen often. It is a huge risk transfer, involving issues of trust and organizational compatibility as well as accuracy and content, but it also relies on compatible system representations and requires data interoperability that might not be present. Better data interoperability is emerging from industry foundation classes (IFCs) and similar projects, but can also be a challenge. Both of these concerns are reduced for design–build firms, where the risk is shared across the entire firm and there is central control of ICT platforms. Incompatible representations tied to the world-view of the user remain an issue—where a designer sees a ten-story curtain wall as a single element, the contractor may see floor-by-floor panelized delivery, installation, and alignment items in their scheduling. Somewhere between design and construction the single conceptual "curtain wall" element must be turned into parts in a "construction process."

This situation may be encouraging the emergence of vertically integrated design–build firms that can control and leverage the data at every stage (Sznewajs and Moore 2016). However, firms in the United States have responded by creating short-term project-centered corporations that share risk and reward among the design and construction team. Continued evolution of new legal and business models seems likely.

Good Data: Finding and Keeping What you Need

From industrial designers to architects and engineers, much of what goes into a design consists of selection or discovery rather than invention: screws, cabinets, hinges, façade systems, electrical cover-plates and switches, mechanical systems, and window glazing are among the items selected from a palette of alternatives. Knowing what is available, what it costs, and what will fit the situation best is a significant challenge in the internet era, with possible suppliers and products scattered worldwide but often left to general wording in a specification or selection by a consultant. Tracking, coordinating, and recording the many decisions that go into the finished building is one of the architect's main jobs. The principal mechanism for doing this remains the traditional "drawings and specifications" of standard contracts, but identifying appropriate products remains a challenge. There are a large number of products available, but there is variation in the quantity and quality of information about the products; descriptions differ in format and content; and there are no integrated software tools to facilitate, coordinate, and document selection (even in this age of the World Wide Web!). These are all factors that limit productivity, forcing designers to repeatedly climb steep product-learning curves or steering them to familiar safe solutions or consultant-selected components. Figure 10.2 illustrates diagrammatically the loss of knowledge that is generally conceded to accompany the handoff from one phase of design and construction to the next—knowledge that might not have to be reconstructed if it were suitably recorded during the design and construction process (Eastman *et al.* 2011).

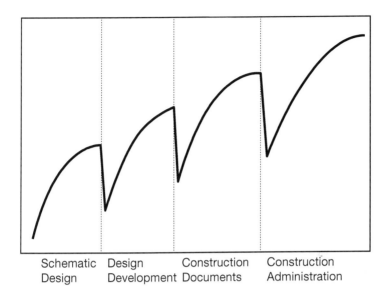

FIGURE 10.2 Motivating the shift to BIM: Knowledge lost during transitions and handoffs.

Standardized digital product descriptions would allow products to be easily imported into differing CAD/BIM systems, tested for connection to products by different vendors, incorporated correctly into engineering analyses, etc. While this has been recognized for many years, especially in Europe, it has proven difficult to achieve the desired level of standardization. Current efforts are focused on IFC representations and their integration into widely used productivity systems.

The problem is not trivial. While good data about products has become more important and the internet has provided a means of access, vendor-supplied product data is not always of a high quality. One consultant noted: "Most manufacturers have data spread out in multiple locations and when we try to enter the information into databases we continually discover errors or conflicts in their information" (Neeley, private communication with the author 2013). A push by large owners and managers, analogous to the push to get designers to employ BIM technologies, may be needed to establish standardized computable representations of components, but design computing research is most likely to identify appropriate strategies.

Access to Data: Managing Focus

Design is not just about acquiring and recording data. Designers must provide data to their clients too, in a series of progressively more detailed views of the project. These are not just data dumps, they are often carefully crafted presentations in which specific questions are addressed and others avoided, not in an effort to deceive the client or sell the project, but in order to guide the project development.

Design is not a linear process. It is a process of successive approximation wherein a candidate solution is refined, revised, and possibly replaced with another. While necessarily cyclical, the goal in each cycle is to move the design forward, so that you are refining and advancing the design, not rehashing. While the design team's focus and attention revisits aspects of the design repeatedly over time, at each point in the cycle of refinement, the designer needs to bring decision-making focus to a particular set of issues, and—equally importantly—ignore others.

Orchestrating and directing this process, avoiding distraction and indecision, is an important skill in the designer's tool-kit. The availability of visualization tools such as photo-real renderings helps tremendously in this process, but the availability and ease of use attached to this technology also raises focus-management issues, especially when clients are given access to a design model that might not be 100 percent ready for prime time, or consultants spend time on analyses using an old model because all data looks the same. Everyone looking at a document set infers which parts are immutable and which are open to change. Strategies for using the technology to minimize these problems by explicitly identifying preliminary ideas or suggestions "under review" within the design are needed. In the 1980s there was a product called Squiggle that added irregularity to straight-line drawings to make them look more tentative and mutable. Something similar might be done for photo-real renderings to make them more "conceptual," such as that used in non-photo-realistic rendering (Winkenbach and Salesin 1994).

Design not only develops in cycles of attention; work is often layered-in-time as well, with multiple overlapping jobs going on in an office over a period of time, possibly disappearing into a file drawer for months at a time as financing, governmental permission, or client planning processes play out. Maintaining appropriate project status information so that a design organization can return to (or pick up) a project weeks or months later is not as easy as preserving the design documents. Formal meeting minutes, request-for-information (RFI), and request-for-proposal (RFP) mechanisms are intended to clarify inter-organizational communications, but offices also utilize personal email archives and individual memories to maintain project state. Whether such data can be efficiently captured on the fly, or retrieved when needed, represents another research opportunity.

Computable Designs

The US government's GSA is one of the largest landlords and construction clients in the United States. To help ensure that proposed buildings meet the requirements of the government agencies funding them, the GSA began requiring "spatial program BIMs" of preliminary building proposals in 2007, with exploratory programs to deepen the use of BIM over time (Hagen *et al.* 2009). Since 2011, the UK government has made an even stronger commitment to use of BIM in the construction industry "with the key objective of: reducing capital cost and the carbon burden from the construction and operation of the built environment by 20%" (UK BIM Task Group 2013).

These efforts are built on the expectation that traditionally graphic information about designs, such as floor plan layouts, have computable aspects (adjacency, area, etc.) that can be extracted automatically and compared to goals. Developing systems to utilize the resulting data and understanding how these government-sponsored technology investments shift the AEC industry, and the ways in which ICT more broadly might further, or hinder, these initiatives, is important to the future of design computing.

Facility Management

The value of data does not end when a building is finally occupied. Large, complex buildings such as hospitals seem to be remodeled continuously, often commencing before the building is fully occupied. Given the complexity of services contained in such facilities, it is not surprising that estimates suggest that maintenance and operations costs of buildings over their lifetime may be triple the initial construction cost (Public Technology, Inc. 1996, citing Romm 1994). The information needed for facilities management (FM) includes not only information about the location of ducts and pipes inside walls (which might be found in as-built drawings), but also information about materials used in construction, equipment manuals, and post-occupancy replacement and maintenance records. This information is intimately tied to the design, and is increasingly part of the package delivered to the building owner during building commissioning. But what form should it take? Including a BIM workstation and project database as an FM deliverable might seem like a good idea, but where do you put it? Who trains the facility operators? Is a BIM program really the right tool? Building operators have different data needs from either designers or contractors. At a GSA workshop attended by the author in 2008, the consensus of the experts attending was that differing information needs mean most projects require three different BIMs: design, construction, and facilities management. Understanding the information lifecycle and information access needs of building operators, emergency responders, and other infrequent building users represents another opportunity or challenge within design computing.

Process, Representation, and Human Organization

The pressure to manage and the need for data in connection with design and construction are creating new opportunities and challenges for practitioners. Chapter 13 will go over some of the ways in which buildings are contributing to and benefiting from emerging data streams. Fine-grained, site-specific information is guiding the form of buildings in new ways. For those drawn to digital concepts and tools there is an opportunity to develop data consulting services for both architects and clients. In an article for *Architect Magazine*, Daniel Davis touches on ways in which clients are demanding data from their designers (Davis 2015).

Anecdotal evidence exists that greater use of 3D visualization in construction-management meetings and drawing sets reduces errors. Digital imagery, from both fixed cameras and mobile devices, is increasingly useful in site-inspection, progress monitoring, and problem resolution.

A lurking problem for any modern offices is file-format obsolescence, in which perfectly good data becomes useless simply because there is no software to read it (or no operating system to run the software, or hardware to run the operating system). New releases of software may employ new representations in order to deliver new functionality. This requires new file formats. The new software will read old files, as well as new, but after a time that old code will be retired. Old files that have not been opened and saved to the new format will become unreadable. This process might take a decade or more to unfold, but most design firms keep archives that span longer periods of time. As the life spans of digital technologies in the workplace grow, data-preservation issues are likely to become more abundant, not just in design practices, but across the world (Sample 2015).

The Decline of Typical Conditions

Traditional designs and design documents recorded decisions about materials, geometry, and the relationships between them, often in the form of *typical* plans, sections, or details. Designers sought to establish a limited "vocabulary" of conditions within the geometry and appearance of the building. Uniform beam spans, column spacing, and material conditions meant efficiency in drawing, efficiency in fabrication, and uniformity in appearance—all features of the "typical" modern building. Buildings that violate the uniformity can become very expensive. The 1991 Seattle Art Museum went roughly 20 percent over budget, at least in part because "not a single steel column runs straight from the basement to the top of the fifth floor [with the result] that hardly any of the individual pieces that make up the building are repeated" (McDermott and Wilson 1991).

In construction, uniformity in formwork, jigs, molds, and dimensions simplifies fabrication. It gives the contractor a chance to learn on one part of the project and use that knowledge repeatedly. While Jørn Utzon's original 1957 design for the Sydney Opera House was relatively simple visually, the iconic shells were initially defined using parabolic curves, which meant varying curvature (and thus varying formwork) across the many rib and panel sections, making them prohibitively expensive to fabricate and forcing that scheme to be abandoned. They were ultimately executed as spherical forms precisely because uniform curvature simplified construction.

Today, design and fabrication systems have become robust enough to support execution of complex designs, such as the Gehry Partners' Guggenheim Museum Bilbao, consisting of many non-uniform parts. While conceived without the computational and data management efficiencies afforded by computing, information technology makes it economically feasible to design, detail, produce, track, deliver, and assemble buildings with such an explosion of atypical parts. The widespread adoption of such technological tools has allowed more and more architects to experiment with gentle undulations in their façades, variation in their glazing or cladding patterns, or geometry in their structural steel. Accurate digital models can feed data to precise digital fabrication machinery that can cut and label individual members and stack them on the truck in the right order, delivered just in time to be placed, while digital survey equipment can ensure correct placement. The result may be atypical, but remains efficient.

Not only are buildings increasingly atypical; modern design practice has become more varied in terms of project location and team composition. In the 1990s academics began exploring distributed or virtual design studios, using digital networks for data exchange and interpersonal communication (Wojtowicz 1995; Hirschberg et al. 1999). Modern high-speed communication enables routine design collaboration across wide areas of the globe, and creates the need to mesh information practices in order to work efficiently. While such global practices are possible, they are found to require a careful selection of collaboration partners (Franken 2005). Systematic study of such collaborative partnerships and their technological foundations would provide another design computing opportunity going forward.

Big Data: Design vs. Form-Finding

Architects and designers are now able to access or generate significant amounts of data in the normal course of a day. Detailed weather, demographic, circulation, and other kinds of data are available for many sites, and both primary tools (CAD or BIM) and secondary tools (analyses, etc.) can be used to produce detailed reports. Paradoxically, two responses are notable: Some designers combine detailed site data and analysis algorithms, often in genetic algorithm or optimization systems, to perform form-finding exercises on the site (Davis 2015), while at the same time an online survey of designers and their use of green building analysis tools found that many who know about such tools opt not to use them in early design (Sterner 2011).

Those who embrace data often seem to want to transduce some quantifiable quality of the site into building form. Bernhard Franken describes a computational process by which project-specific fields of physical and notional ("not strictly physical") forces act on geometric forms to produce a "master form" from which the design documents and detailing are derived (Franken 2005).

Using Simulation

In light of broad awareness regarding the impact of early decisions on design performance, it is surprising that greater use is not made of energy simulation during schematic design. The online survey mentioned above revealed that while roughly half of respondents indicated familiarity with green building analysis tools and frequent participation in green building projects, very slight actual use was reported of such tools during design, even among those who considered themselves experts in green building design. The report identified dissatisfaction with "ease of use" and "time required" as major challenges, leading the report's author to conclude, "most practitioners are presumably relying on rules of thumb, prescriptive checklists, and/or consultants." Similar results have also been reported by others, who also found "Architects today rarely use whole building energy analysis to inform their early design process" (Smith *et al.* 2011). They further identified "understanding the results and using them in presentations" as inhibitors. For these users, data, in the form of analysis results, seems to be more problem than opportunity.

Why might this be so? The truth is that building a good simulation model is something of an art and not simply a translation of building geometry—"There [are] too many opportunities to fail with an energy model that [is] derived from a detailed building model" (Smith *et al.* 2011). Simulation is not routine. It is not common. It is a realm of professional practice that is being ceded to an evolving collection of expert consultants, which locates simulation adjacent to design practice, not central to it.

Different authors have noted that the domain of design has grown narrower over the years as architects "spin off" decisions to related specialists—structural engineers, materials consultants, etc. Of this "retreat to the protected core" John Archea wrote

> it seems that whenever a body of knowledge develop[s] to the point that the effects of a specific building attribute or system could be fully understood prior to and independently of the initiation of the design process, architects have redefined their profession to exclude responsibilities for achieving those effects.
>
> *(Archea 1987)*

Whether this behavior is a response to increasing complexity, a response to risk, or a means of preserving identity, it raises sociological and psychological questions about how best to combine design and computing, and how best to inform an integrative design process.

Research, Development, and Continuous Learning

In light of the above, one of the interesting changes in the last few years has been the re-emergence of in-house research groups within architecture firms. In the 1960s and 1970s a number of firms had active software development efforts, including Skidmore, Owings & Merrill (SOM) and Hellmuth, Obata & Kassabaum (now HOK) in the United States, and Gollins Melvin Ward (now GMW Architects) in the United Kingdom. These projects were largely abandoned or sold as the rising tide of micro-computer CAD washed over the industry in the 1980s, but offices in the early years of this century have witnessed a rebirth of support for in-house research, development, and training in the face of rapid technological change in the industry, including the growth in data exchange options and the advance of end-user programming or scripting functionality in commercial software systems.

This re-awakening coincides with an increased emphasis on STEM (science, technology, engineering, and math) studies and "code" across the broader culture, and increased access to and awareness of open-source software-sharing communities, to the point where—concerning technology use—"many students and practitioners have reached a certain fluency with the existing tool sets and strive to distinguish themselves through customized design processes" (Kilian 2006).

The addition of continuing-education requirements to architectural licensing also energizes the professional and business case that can be built for in-house training and sharing regarding design technology, reinforcing and broadening the tradition of sharing and reflection on design topics that has been practiced in many offices over the years. The net result is a proliferation of both external and internal blogs, knowledge communities, corporate universities, and conference opportunities. Developing and supporting the knowledge and culture necessary to identify and harness these emerging best practices provides another avenue for design computing practitioners.

Summary

In this chapter we have seen opportunities in the routine praxis of design and construction for those who study design computing to contribute not only to the efficient and knowledgeable operation of today's technology, but to engage in activities as disparate as creating new software, re-forming the legal foundations of the industry, and shaping the internal social and external knowledge-sharing communities of firms and communities. At a deeper level are long-term challenges regarding the best way to use data in design practice when each analysis is just one source of information in a complex multi-faceted juggling act. Design is about integration and coordination, where the details are often left to consultants; yet those details have significant impact on the overall project and may need to be revisited at a later date. Capturing, archiving, curating, searching, and accessing information in the design process is increasingly valuable. Modern practice relies heavily on digital technology, but we are still searching for the best fit between creative design and the devilish details.

Suggested Reading

CURT. 2010. UP-1203 BIM Implementation: An owner's guide to getting started. Construction Users' Round Table. April 20.

Davis, Daniel. 2015. How big data is transforming architecture. *Architect Magazine* (April 23) www.architectmagazine.com/technology/how-big-data-is-transforming-architecture_o.

GSA. 2007. *GSA BIM Guide Series 01 (ver 0.6)*. US General Services Administration. www.gsa.gov/bim

Hagen, Stephen, Peggy Ho, and Charles Matta. 2009. BIM: The GSA story. *Journal of Building Information Modeling* (Spring): 28–29.

References

Archea, John. 1987. Puzzle-making: What architects do when no one is looking, in *Computability of Design*. Edited by Y. Kalay, 37–52. New York, NY: John Wiley.

Campbell, Dace. 2004. Building information modeling in design–build. Talk presented at the University of Washington, December 2.

CURT. 2010. UP-1203 BIM Implementation: An owner's guide to getting started. Construction Users' Round Table. April 20.

Davis, Daniel. 2013. Modelled on software engineering: Flexible parametric models in the practice of architecture. Unpublished PhD dissertation, RMIT University.

Davis, Daniel. 2015. How big data is transforming architecture. *Architect Magazine* (April 23) www.architectmagazine.com/technology/how-big-data-is-transforming-architecture_o

Eastman, Charles M. and Rafael Sacks. 2008. Relative productivity in the AEC industries in the United States for on-site and off-site activities. *Journal of Construction Engineering Management* 134 (7): 517–526.

Eastman, Chuck, Paul Teicholz, Rafael Sacks, and Kathleen Liston. 2011. *BIM handbook: A guide to building information modelling for owners, managers, designers, engineers and contractors*. Hoboken, NJ: John Wiley

Franken, Bernhard. 2005. Real as data, in *Architecture in the digital age: Design and manufacturing*. Edited by Branko Kolarevic, 123–138. New York, NY: Taylor & Francis.

Graves, Michael. 2012. Architecture and the lost art of drawing. *New York Times*, September 1.

GSA. 2007. *GSA BIM Guide Series 01 (ver 0.6)*. US General Services Administration. www.gsa.gov/bim

Hagen, Stephen, Peggy Ho, and Charles Matta. 2009. BIM: The GSA story. *Journal of Building Information Modeling* (Spring): 28–29.

Hancock, Lillian. 2013. Visualizing identity: Perspectives on the influences of digital representation in architectural practice and education. Unpublished Master's thesis, University of Washington.

Hirschberg, U., G. Schmitt, D. Kurmann, B. Kolarevic, B. R. Johnson, and D. Donath. 1999. The 24 hour design cycle: An experiment in design collaboration over the Internet. *Proceedings of CAADRIA '99, the fourth conference on computer aided architectural design research in Asia*, 181–190.

Kilian, Axel. 2006. Design innovation through constraint modeling. *International Journal of Architectural Computing* 4 (1): 88–105.

Kolarevic, Branko and Ali Malkawi (eds.). 2005. *Performative architecture: Beyond instrumentality*. New York, NY: Spon Press.

M.A. Mortenson. 2005. Case study: Denver Art Museum. www.mortenson.com/approach/virtual-design-construction/~/media/files/pdfs/denver-art-museum.ashx.

McDermott, Terry, and Duff Wilson. 1991. Museum officials feared, discussed big cost overruns. *Seattle Times*, July 14.

Public Technology, Inc. 1996. *Sustainable building technical manual: Green building design, construction, and operation*. Washington, DC: Public Technology, Inc.

Robbins, Edward. 1997. *Why architects draw*. Cambridge, MA: MIT Press.

Romm, Joseph J. 1994. *Lean and clean management*. New York, NY: Kodansha International.

Sample, Ian. 2015. Google boss warns of "forgotten century" with email and photos at risk. *Guardian*, February 13. www.theguardian.com/technology/2015/feb/13/google-boss-warns-forgotten-century-email-photos-vint-cerf

Scheer, David Ross. 2014. *The death of drawing: Architecture in the age of simulation*. New York: Routledge.

Smith, Lillian, Kyle Bernhardt, and Matthew Jezyk. 2011. Automated energy model creation for conceptual design, in *Proceedings of SimAUD '11 symposium on simulation for architecture and urban design*. Edited by Ramtin Attar, 13–20. Society for Modeling and Simulation International.

Sterner, Carl S. 2011. Architecture software survey: Results & analysis. Sterner Design. www.carlsterner.com/research/2011_architecture_software_survey.shtml

Sznewajs, Timothy and Brian Moore. 2016. Construction industry mega trends emerging from the recession: What every CFM needs to know. Construction Financial Management Association. www.cfma.org/content.cfm?ItemNumber=2533.

Teicholz, Paul. 2013. Labor-productivity declines in the construction industry: Causes and remedies (another look). *AECbytes Viewpoint* 67 (March 14).

Teicholz, P., P Goodrum, and C Haas. 2001. U.S. construction labor productivity trends, 1970–1998. *Journal of Construction Engineering Management* 127 (5): 427–429.

UK BIM Task Group. 2013. Home page. www.bimtaskgroup.org.

Winkenbach, Georges, and David Salesin. 1994. Computer-generated pen-and-ink illustration, in *Proceedings of Siggraph'94*, 91–100. New York, NY: ACM.

Wojtowicz, Jerzy. 1995. *Virtual design studio*. Hong Kong: Hong Kong University.

11
EXPERTISE
Challenges and Opportunities

> An expert is someone who knows some of the worst mistakes that can be made in his subject, and how to avoid them.
>
> *Werner Heisenberg (1971)*

The built environment touches almost all domains of human knowledge: history (symbolism and culture), physical sciences (behavior of materials in complex configurations), biology (response to materials and environment), psychology (behavior of people in various settings), engineering (distribution of energy and support), and a host of other disparate fields of study. These are each subjects in which individuals earn doctoral degrees, so it is not surprising that those responsible for designing new environments do not possess all the required expertise. Their expertise lies in design—consulting, integrating, and coordinating the great many details connected to the expertise of others.

In the face of all the knowledge needed to design and construct anything, acting can be a challenge. Reducing expert decision-making to best practices, rules of thumb, prescriptive code guidelines, and some number of advisory consultations enables the designer to see past the trees to the forest and make needed decisions.

Most designers acknowledge that discovery and continuous application of expert advice beginning early in the design process might well save time and money over solutions based on rules-of-thumb, but also recognize, as we saw in Chapter 8, that current design and simulation tools are not necessarily easily applied at every stage from schematic design through design development and documentation, and that too much detail can be a distraction and hindrance to decision-making as well. Further, the fragmented structure of the industry may require incomplete specification of decisions in order to create price variation in the bidding process.

This chapter will look at the challenges and opportunities that accompany the application of expert knowledge and judgment within the design process. These questions necessarily interact with the topics of cognition (Chapter 7) and representation (Chapter 8).

Architects are Generalists

Architects are generalists; they rely to a great extent on the expertise of others to both design and build projects, and yet they must oversee the myriad decisions that are required during design. They may consult experts for advice on everything from code interpretations to structural systems to fireproofing products to lighting strategies to sustainability and siting. Some questions are simple knowledge questions, but others require expert, numeric analysis of a design proposal. Further, the specificity, or level of abstraction, of the questions can change depending on the phase of the project, so some come up repeatedly as scale and detail change.

When analysis is needed, projects may be evaluated in any of a large number of very different dimensions: including aesthetics, construction and operating cost, structural integrity, energy use, lighting quality (both day lighting and electric lighting), and egress (emergency exiting). Some analyses are mandated by architecture's core professional commitment to life-safety, while some are mandated by governmental, client, or financing organizations. It is not necessary to evaluate every building in every dimension, and many evaluations (e.g., egress) produce simple binary assessments—the project either "meets code" or it doesn't. Even where variability is expected (energy use), prescriptive component performance specifications are often used in place of an analysis that responds to the actual building geometry, materials, and site. Finally, analyses done during schematic design may be quite simplified compared to analysis performed during design development or construction document preparation.

The market dominance of a small number of productivity tools (CAD or BIM) may leave the impression that there is a single common representation of built environments. Though this is the direction in which software seems to be evolving, it is still far from true. As discussed in the chapter on representations, the information needed for one analysis is often quite different from that needed for another (e.g., early cost estimates might depend only on floor area and building type, design-development estimates might require precise area takeoffs and material/quality selections, while final construction estimates might add in localized cost data, and perhaps even seasonal weather adjustments). In the past, a separate representation (data set) would often be prepared for each analysis, featuring only the information required for that analysis. As the project advanced, the data set could be edited to reflect design changes and the analysis could be re-run, or a new data set could be generated from the then-current set of design documents.

Different analyses are often performed with both different spatial and temporal resolution; a simple daylight penetration analysis may consider only the noon position of the sun at the extremes of solar position (summer and winter solstice) plus the equinoxes; a mid-level analysis might check the twenty-first of each month; and a detailed analysis might consider hourly solar position in concert with occupant activity schedules, etc. Rough energy analysis might ask only for building area and occupancy. Early structural analysis will consider general structural system, occupancy, and column spacing, while advanced evaluations require precise building geometry, plus local seismic and wind-loading data in addition to soils and code information. However, even the advanced structural analysis will probably ignore the presence of non-load-bearing walls, though these may well define the boundaries of heating system zones, rentable areas, etc. Decisions about what is important rely on expert knowledge in the various consultant subjects.

Asking the Right Questions at the Right Time

While serving as generalists, architects must function as the integrators and coordinators of outside expertise provided by collaborators and consultants. They know that a great many questions need to be answered and that getting them addressed in the right order will save time and money. They know that the cyclic nature of design will cause them to revisit decisions, examining them from different frames of reference and at different scales as the project develops. Because decisions interlock and cascade (resonance or synergy between related decisions is often seen as a positive feature of a good concept) they cannot simply be ticked off some checklist. Design is both a process of asking and answering questions, using design propositions to explore and test possible futures and various analysis tools to evaluate the results. The nature of that evaluation is not a given; it depends on the question that has been asked. While it might be ideal if you could compare current results with earlier ones, testing design moves with "before" and "after" results, the time and money expense of doing this remains high enough that designers rarely work this way. Designers have (or can easily find online) simple rules of thumb for many standard building system elements. When it is necessary to consult an expert, the question asked is less likely to be "How much does this recent move change my performance metric?" It is more likely to be "What's the annual energy demand of this configuration?" or "What level of daylight autonomy can I achieve with this façade?"

Within the disciplinary silos of design and construction—architecture, materials science, mechanical engineering, structural engineering, day-lighting, etc.—there are well-tested analysis and simulation tools able to address different levels of detail and kinds of question. However,

> it should be realized that the building industry has characteristics that make the development of a "building product model" a huge undertaking.... Major obstacles are the scale and diversity of the industry and the "service nature" of the partnerships within it.
>
> *(Augenbroe 2003)*

So, while there are suitable analysis tools and abundant expertise, the scenario in which off-the-shelf digital technology would be as easy to apply as a spell-checker does not emerge. Instead, "building assessment scenarios typically contain simulation tasks that … require skilled modeling and engineering judgment by their performers" (Augenbroe 2003). Automated production of feedback that is appropriate to the current phase of design and the designer's shifting focus remains a serious challenge.

Drawings are Never Complete

While it may be obvious that "back of the envelope" sketches and even schematic design drawings are incomplete and rely heavily on numerous assumptions about material choice, dimensions, etc., it is also true that finished construction document sets are incomplete in various ways. First, while they may well conform to industry expectations and legal standards, they do not show every aspect of the finished project. Some details are left to the individual trades to sort out, with shop drawings sometimes being generated by the sub-contractor to show the precise configuration proposed. Other details are covered by general industry standards set by trade groups. Still others simply exist as best practices within trades.

Expertise and the Framing Problem

Every consultant sees different critical information in the building. The elevator consultant sees wait times, floor occupancies, weight limits, building height, and scheduling, while the lighting consultant sees window geometry, glazing material, finishes, tasks, and energy budgets. Each of these conceptual "frames" can be drawn around the same building, but not exactly the same data. Most require the addition of some specific information that derives from their point of view. The architect may select a floor material based on color, wear, and cost, but the lighting consultant wants to know the material's reflectivity in the visible spectrum, and the sustainability consultant is concerned about the source, embodied energy, and transport of the material. All this information is available, somewhere, but probably *not* in a BIM model of the building, much less in the CAD drawings. Nonetheless, decisions *will* be made based on real or assumed values. Capturing and recording values as they become known is a challenge, but understanding what is assumed in your tools might be even more important. Filling in that information with appropriate informed (expert) choices is one of the roles of the consultant. Doing it as part of their analysis is common, and that's a problem if the analysis needs to respond to, or propose, design changes.

Consultants and the Round-Trip Problem

Figure 11.1 illustrates what is sometimes referred to as the "round-trip problem" by zooming in on the portion of the design cycle where outside expertise is often

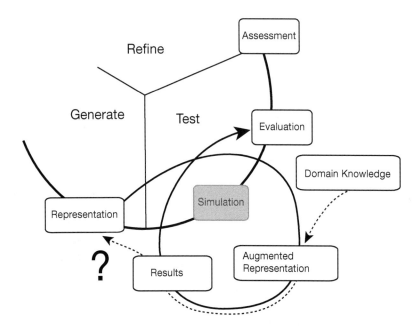

FIGURE 11.1 The Round-Trip Problem – integrating expert domain knowledge into the design model in preparation for further design cycles.

sought—simulation of performance. When an expert is brought in to perform analysis on the design they will receive data from the designer, but often need to bring additional domain knowledge or detail to the model, as indicated by the dashed line. Further, their field of expertise will very likely *require* a different framing of the design. In the end, their analysis model will *resemble* the design model, but not match it exactly. For example, they may have overlaid a set of thermal or acoustic zones on the building plan, zones not represented explicitly in the design representation of the architect. The results of their analysis, taking into account the added information, will feed back to the design evaluation, but their model, which is separate from the design model, will exist outside the loop. Capturing and incorporating into the design model all the specific changes required to enable their analysis is rarely done, or even attempted, as indicated by the question-mark. Reintegrating results into the building model in such a way that consistency and integrity of the model are maintained is problematic in practice and difficult to develop (Augenbroe 2003). This difficulty delivering a "round trip" for detailed analysis data remains one of the roadblocks to efficient use of building models with consultants, and becomes increasingly important as workflows begin to incorporate expert knowledge earlier and execute multiple cycles of simulation (Kensek 2014a, 2014b).

Currently, if a second analysis is required and the same consultant is used, they may simply update their original model, replacing or updating it as appropriate, depending on the number and extent of design changes, and re-run their analysis.

This is neither a very fluid exploration (from the point of view of the architect) nor an inexpensive one, which is why much design work simply begins with heuristics (rules of thumb) and uses analysis just once at the end, to verify performance.

Growth and Change

Of course, design organizations seek to build in-house expertise as well as consult outside expertise. And individuals gain experience that they then tap during future projects, but building institutional memory requires a process to extract and collect assessments of past work—best practice details or assemblies. Offices keep archival copies of their design documents for legal reasons, often in the form of CAD files, so one fairly common practice is to browse the archives for similar projects and then extract decisions from the old project for use in new ones. The difficulty here is the absence of feedback loops—the CAD manager at one prominent Seattle firm once complained about the difficulty of "killing off" a particular roofing detail that kept cropping up in projects even though it had proven problematic in the field. Explicit project-reflection or knowledge-management systems remain a work in progress.

While designers tend to be generalists, there is increasing need for data manipulation skills connected to expertise, and more young designers who are building reputations around data visualization, parametric design, and related questions. This shift has moved data visualization and manipulation much closer to the core of design.

Another change that is occurring is in the level of expertise that designers are expected to bring to and exercise in the design process. In the twentieth century, as building-wide heating, lighting, and ventilating became common and high-rise steel construction was widely used, analysis tended to focus on establishing minimum threshold levels of light, heat, strength, etc. In the twenty-first century, as we become more sensitive to the waste associated with over-design, there is a trend toward more sophisticated analysis employing higher-resolution data and more complex models, further isolating such work from routine design activity. Individuals and firms that specialize in the relationship of data, designers, and expertise are increasingly common.

Harvesting and Feeding Leaf-Node Knowledge

Architects are not master masons or steelworkers; they rely on the expertise of laborers or supervisors to build projects. In the flow of knowledge from the architectural trunk through the various branches of contracts and sub-contracts these workers are the "leaf nodes." They are the ultimate eyes, ears, and hands of the design team; they know how to nail, drill, bolt, weld, cut, and assemble. And while traditional construction documents do not tell contractors *how* to execute a project (that's left to their expertise), sometimes the expertise out at the leaf would help the designer make a better choice. Unfortunately, the economics of bringing such fabrication expertise into the architectural design process, where each building is a "one-off" design, is daunting.

To illustrate the potential value of such information, consider the Boeing Company's "Move to the Lake" project (NBBJ 2016). Designed by NBBJ, the project combined factory and office facilities into one co-located space, dramatically reducing the overhead of moving people and information from manufacturing to engineering or vice versa. They not only won praise and awards for locating the design engineers' offices adjacent to the actual plane production area, the time to produce a plane was cut by 50 percent in the process. This architectural approach dramatically changed costs. The potential for information technology to do something similar within environmental design is compelling, if challenging. Whether such tactics can be put to work in the United States building construction industry, where most construction is done by constantly shifting constellations of small firms, remains unclear.

If changing information systems might not make expertise more available to designers, it is fairly clear that changing contractual relationships can. Alternative models of contracting, including integrated project delivery (IPD) and design–build–operate (DBO) models, are gaining popularity, appealing in part because they offer the greatest synergy between current information technology and other users of the information. By bringing fabrication knowledge into the design process sooner, the MacLeamy Curve may actually begin to pay off in terms of impact on costs.

The benefits are not limited to the harvesting of leaf-node knowledge; more and better information can make the contractor's job easier as well. M.A. Mortenson has adopted a strategy of bringing 3D visualizations to their daily construction-management meetings, after seeing a reduction in errors related to placing steel embeds during construction of building cores when the folks in the field had a clearer understanding of the ultimate configuration of the building structure.

Summary

The design process invariably leads designers to engage with expertise. Most design evaluation depends on digital models, each of which requires a uniquely flavored blend of information from the design data, depending on design phase and nature of the question(s) being asked. Answers should often be fed back into the design for the next iteration. Bringing efficiency to the process of tapping expertise, including constructors, is a serious challenge as it touches on relatively fluid representation and process needs. Evidence from other industries suggests how valuable such information might be and changes in the AEC industry are occurring as different industry players attempt to position themselves to best use the tools currently available.

Use of 2D drawings to represent designs creates "unstructured" data (Augenbroe 2003), which means that experts familiar with the design are needed to interpret drawings and prepare data for analyses. Even where 3D BIM models are available,

This may be as easy as comparing an existing metric (e.g., percentage of envelope that is glazed) against a legislated (code) value, but it may also involve a more complex compliance (e.g., a passive solar design balancing glazing with thermal storage, or implementing a novel automated shading system) in which case *before* and *after* performance may need to be compared through simulation. In the context of our notion of a design space, these amount to *localized* explorations of design space.

As discussed in the context of detecting solutions to problems, designs are complex phenomena, simultaneously serving goals that range from occupant safety and comfort to civic or corporate image, to designer branding. Often, no single computable metric gathers and reflects all these goals, so it is hard to numerically assess and compare designs side by side. However, if we restrict our interest to a narrower goal, like improving energy consumption, we can define a fitness function that does let us compare. Sadly, we are rarely able to perform the kinds of symbolic manipulations used in calculus to find minimum and maximum values. Instead, we have to test each possible solution. This is the land of the traveling salesman, who we met first in Chapter 1, a land that cripples computer systems.

Computers simply aren't fast enough. Conceptually, we should be able to perform an *exhaustive search* of a solution space, testing every possible solution and keeping track of the one with the best performance. Aside from the problem of ineffable qualities not explicitly represented in the state-space, the *combinatorial explosion* problem means that state-spaces are usually very large, even "vast," and contain very few good solution candidates (Woodbury and Burrow 2006). Further, as discussed previously, there is no reason that these candidates would be clustered. A purely random selection from the space has a very low probability of being a good fit.

However, by comparing several candidate solutions, perhaps randomly generated, and the relative values of their fitness functions, we might be able to identify an appropriate *change* to the design, one that will improve the result. If this can be done reliably and repeatedly, we have a way to move towards that unknown optimum through a process known as *hill-climbing*, described below.

Alternatively, building on Darwin's theory of natural selection, *genetic algorithms* apply a strategy which is neither an exhaustive search nor completely random. They use the state-space axes or dimensions to define *genes* for the design. Corresponding to a state in state-space, the defining values of a particular design are called its *genome*. Genetic algorithms begin by producing a set of random candidate designs (a *population*). These are then tested with the fitness function to identify the strongest members. The fittest members of the population are then *bred*, combining genes of one relatively fit design with those of another while adding a little random *mutation* now and then, to produce a new population of possible solutions. Over many generations (possibly thousands), under the influence of the fitness function, the survivors score better and better on the fitness test.

Objective Functions

If there were a process to follow to get directly from identification of a problem to the solution, design would be simpler; but this is rarely the case. In fact, it is *iterative*, and each cycle will produce a possible solution. Unfortunately there is no guarantee that each cycle will produce an *improvement* in the design—sometimes it gets worse. Given two alternative designs, we need to be able to tell which one is better. This is the purpose of the *fitness*, *utility*, or *objective* function, an equation that allows a design to be scored numerically.

To understand the concept of an objective function, consider the following. Floor plan layouts cannot superimpose two rooms on the same area of the plan, but proximity is often a desirable condition (think dining rooms and kitchens, kitchens and entries, nurse's stations and pharmacy stores, assembly-step A and assembly-step B, etc.). One aspect of architectural programming is to establish the relative importance of different proximities for the client. Given a floor plan, this information can be used to score the design in terms of how well the overall plan satisfies the different proximity preferences. The proximity score, or circulation cost, serves as the design's objective function. A good design is one in which that cost is minimized.

The objective function can tell you which of two designs is *better*, but it cannot tell you how to produce the *best* layout, and most designers would assert that there is no *single* objective function, or that good designs frequently have to compromise some aspect of the building's function. One area of active research in design computing is evaluation of designs using multiple criteria, a field made possible by application of computer technology.

Multi-criteria optimization, also referred to as multi-objective optimization or Pareto optimization, is used to address problems with multiple objective functions. Candidate designs, called feasible solutions, are subjected to evaluation using a set of objective functions. Recognizing that these evaluations interact, the approach doesn't optimize each dimension of the problem, but seeks to identify those feasible solutions that optimize each objective function without damaging the results of the others. This approach usually identifies multiple candidate designs within the design space. As a collection, they will delineate a line or surface within the solution space. This edge is referred to as the *Pareto front*, named after the Italian economist Vilfredo Pareto. It is reasonable to suppose that all points along the front satisfy the multi-objective optimization goal, though the designer is free to choose between them.

Solution by Inspection

In early design, neither the existence nor the character of final solutions are evident to the designer. The situation might be likened to that of a mountain climber determined to conquer the highest mountain in the world. If presented with a sorted list of every peak in the world over 5000 m (15,000 feet) in height, it would be a simple matter of reading the top item off the list and packing your gear. But if you were standing in a valley in the Himalayas, perhaps shrouded in fog, it is harder to pick out the right target. Indeed, without a "God's eye view" providing more information than is generally available, even when you've climbed a peak you may not know if there are higher ones hiding in the clouds. Solution by inspection is rarely available to a designer. Only by ascending each can their heights be measured and compared. There is a complex terrain and we know what it means to be higher, but it is unclear how to get from the valley to the peaks, so we *move* about in the design space, and we *assess* the impacts of our moves as we go. When we think we've reached a destination, we *stop*.

State-Space Neighborhoods

Our view of the design cycle must include generating alternative solutions by manipulating a representational scheme, comparing them via one or more objective functions, keeping the best candidate at each step, and comparing results from multiple evaluations to suggest the next move. Each alternative solution is a state in the design space of the problem, an elevation at a location in our climber's valley. The systematic generation of all possible solutions and their testing against selected performance criteria, while theoretically possible, would waste much time. As we discussed previously, the size of the design space for all but the simplest problems renders this strategy unworkable in the real world.

Inefficient but reliable might be acceptable, but one of the characteristics of architectural design space in general is its vastness (Woodbury and Burrow 2006), and the scarcity of viable solutions. While the size of the search space makes exhaustive search impossible and the paucity of viable solutions makes random testing equally unworkable, there may be ways to conduct intelligent search. Perhaps we can generate solutions from first principles, tap human collaborators (or history) for potential solutions, or explore incremental modification of prior candidates. In the past the search for efficient ways to use computational power to solve problems or improve partial solutions has taken various forms.

One strategy works like this: Even if our hypothetical mountain climber is completely fog-bound and cannot see the peaks and thus cannot select a destination by inspection, she can still climb. She can take a test step in each of several directions and then move in the direction of the most rapid ascent. This *hill climbing* strategy (yes, that's what it is called) doesn't require a symbolic representation of the design-space landscape, only the ability to change the design state and apply the objective function. After every step is taken, the process repeats. When you reach a point where every step you take leads down, you are at the peak, even in the fog. Hill climbing does rest on the assumption that there are no discontinuities (cliffs?) in the design space. And it recognizes that you might find a *local* maximum, which is why these approaches usually involve a meta-cycle with a number of random start points. If you keep finding yourself on top of Mt. Everest, the thinking goes, your confidence in it as "the" highest peak goes up.

Hill climbing relies on the objective function being independent of the search. That is, it doesn't work well on boggy ground where every step is a step up, until you take it and your weight begins to change the landscape as you sink into the bog. These are the wicked problems.

Selecting Design Moves

Each transformation of the design, or *move*, involves selection from a long list of possible actions. In geometry editors such as most current CAD/BIM software, the selection is left entirely in the hands of the designer. In rule-based systems such as shape grammars, the viable options may be presented to the designer for a selection. In any case the move brings the design to a new and different state. Tools to advise or support the designer in making these decisions would be immensely useful if they worked reliably, but may have to wait for AI. In the meantime, the goal of conducting an efficient search of design-space remains. Two promising alternatives have been developed in recent years: directed search and genetic algorithms.

Directed Search

Disregarding for a moment the quote from Denys Lasdun with which we began this chapter, most clients have some idea of what they want. Even if they don't, representations such as shape grammars have embedded cultural or stylistic framing that limits the design-space significantly. The implicit constriction implied by such tools makes possible various kinds of directed search, in which the designer selects directions at crucial points in the decision tree, but software produces and possibly evaluates candidate decisions for exploration.

Computer-science researchers use machine-learning algorithms to identify patterns in complex human endeavors. This is the realm of "big data" and AI. In a more organic form, developers of apartment buildings or condominiums often have a finely honed sense of what features sell. While still relatively rare in academic architecture, researchers at Stanford University have used machine-learning techniques to analyze sets of floor plans and generate candidate plans in relation to new architectural design problems (Andia and Spiegelhalter 2015), and shape-grammar systems such as those developed at Carnegie Mellon University present the operator with sets of alternatives at each iteration of the design, keeping the design "within bounds" while allowing human choice (Heisserman 1991).

Genetic Algorithms

Another promising approach, first developed by John Holland, is the *genetic algorithm* (GA) (Mitchell 1996). Following the classical problem-solving strategy of working through analogy, equating problem solving with success in the biological world, Holland and his colleagues compared design space axes to biological genes, whose values and interactions determine traits, which interact with the environment in such a way that the individual thrives or dies. Those who live are able to reproduce, combining their genes with others, creating hybrid offspring with new traits. Over time, those genes and traits that help an individual thrive appear in more and more of the population, meaning that they represent a solution to the problem posed by the environment. Further, mutation can radically change an individual, increasing or decreasing their viability.

In biological evolution, natural selection culls weak individuals before they can reproduce, a process in which attributes of one parent are combined (called "crossover") with attributes of another, producing offspring with traits from both parents. Mutation assures that the population acquires new traits from time to time.

In GAs, the design problem and its criteria define the environment. Solution attributes (dimensions, configurations, adjacencies, etc.) form the genes. A population of randomly generated individuals (designs) is tested against the selection (objective) function. Some individuals survive. Their genes are combined in a mix-and-match fashion to make a new generation of designs. A few random mutations are thrown in, and the new generation is evaluated. After many generations (sometimes thousands) of selection, crossover, and mutation, a combination of traits tends to emerge and dominate the population.

This approach does not presume the existence of a correct representation or best design, nor can it guarantee that solutions are optimal. It allows interacting-genes to be tried out in various combinations. It jumps to entirely new parts of the state-space through mutation and it explores neighborhoods of variation through crossover. In a sense, it does hill-climbing on a very broad scale, using pretty simple operations and lots of computer time. It can produce unexpected results, as designers of a NASA X-band antenna found when they applied a GA to the problem (Hornby et al. 2011).

Evaluation (Comparing Apples and Oranges)

Once a design move has been made, the design cycle calls for re-evaluation of the design. In practice such re-evaluations are often delayed until a number of changes have accumulated because of the time required to perform an evaluation. Evaluations are usually indirect, since the design artifact does not yet exist, and they may also use quite different approaches. The following paragraphs illustrate several approaches to a "how much energy does it use?" question.

Direct Measurement

To compute energy consumption of an existing building we need only meter the gas and electricity being delivered to the building over a period of time, and perhaps survey the occupants regarding activities in the space, appliance use, etc. Buildings not-yet-built present a more complicated problem.

Measurement by Analogy

Great swathes of construction follow similar standards due to concurrent construction, building code constraints, or cultural bias on the part of builders or buyers. A low-fidelity evaluation might well be created by looking up the measured impact of similar moves on similar construction. This means that one way to estimate the future energy use is to monitor or measure historical use of similar existing buildings.

Simulation

Significant science has been done on material physics and the study of heat transfer. Nonetheless, objective answers to "how much energy?" are not simple to calculate. Computing the thermal behavior of portions of a design under unvarying ("steady state") conditions can often be done with equations derived from laws of physics, but in the real world conditions vary over time with changing weather, sunshine, occupant behavior, equipment use, etc. Were these changes to follow simple functions (sine, cosine, etc.) we might be able to handle them with calculus, but they do not. Results depend on the simulated weather and usage, but what are the appropriate conditions for a computation? Designing for average weather ignores or minimizes the discomfort occupants will feel during unusual cold snaps.

While all these factors go into overall performance assessment, comparative assessment mostly means subjecting before and after models to the same analysis process.

Prototyping

In certain circumstances where performance is of paramount concern and there are neither historical examples or believable science models, it may simply be necessary to build a prototype, subject it to wind, rain, people, or whatever, and see how it performs. Careful recordkeeping can provide appropriate data to include in future simulations.

Once the individual analysis results are available, the designer must consider (evaluate) whether they require additional changes in the design. In some cases this will be straightforward—steel columns or beam sizes might need to change if wind loads increase. In other cases it might be very difficult—changing from a steel structure to a concrete structure might mean changing column spacing, changing how building services fit within the structure, etc. Exiting problems might require addition of a stairwell. Too little or too much daylight might require reconfiguration of the façade. Of course, each of these direct responses is likely to interact with other analysis results; money saved through additional daylight harvesting might be lost through extra heating or cooling requirements. It is impossible to get a perfect result across all criteria, which means judgment is required as to the overall acceptability of the result, and what design move might be executed if a change is needed.

Stopping Rules

It is commonly observed that designs are never complete, that there are always ways they might be improved, but the designers just run out of budget or time. This is a side-effect of the orthogonality of problem dimensions and the lack of analytic representations. In purely human design environments, work is usually organized in ways that move from big decisions with substantial cost and human impact to smaller and smaller decisions that gradually resolve problems with the design. Similarly, in a digital design environment such as an optimization program it is necessary to establish some sort of *stopping rule*. Search algorithms may asymptotically approach "best," but they will not arrive. At some point it must be "good enough," or what Simon referred to as "satisficing" (Simon 1969).

The Creative Machine

No discussion of the production of design solutions would be complete without consideration of the potential for software to take over the task of design. This is not a simple topic. Designers invest significant effort in searching, organizing, tracking, and manipulating sizable amounts of data, often in very predictable ways. Given a few simplifying assumptions it should be possible to automate much of that. In fact, Skidmore Owens and Merrill (SOM) developed the *Building Optimization Program* (BOP) in the 1960s using minimization of costs across structure, exterior skin, mechanical system and elevators as the central computation. The result, "a kind of design prosthetic to enhance and broaden the scope of an architect's ability" (Lui 2015), generated the bland commercial boxes of that era and was still used "as late as 1990" (Lui 2015). While cost-effective, these were not creative designs. Still, this wasn't what killed BOP. Ultimately, BOP, like many other academic and corporate programs, was wiped out by the rising tide of personal computers and 2D CAD.

The rigid, inflexible graphics of early 2D CAD systems, which could be scaled proportionately but not altered parametrically, have now given way to architectural BIM objects (doors, windows, floors, stairs, roofs). These parameterized objects might provide a foundation for directed search and GAs to be applied in an explicitly architectural context, but they are currently limited to human-directed model editing.

This situation has exacerbated the division between design (human) and representation (computer). Design is creative; computers provide a medium of design expression. Nonetheless, creativity research has surged in recent years, in fields close to and far from traditional design fields. Computing and business researchers join neuroscience and design investigators seeking to develop scientific approaches to questions of design and creativity (Dorst 2011). Yet, skepticism about the machine's ability to contribute to a creative outcome persists. One reason is "Lady Lovelace's objection" (Mitchell 1977).

Lady Lovelace's Objection

Most human cultures value *thinking* and *creativity* as core human traits. The impossibility of getting from the known to the unknown was even addressed by Plato in *Meno's Paradox* (Meno 80d-e). Debate as to the ability of computing machines to do either began again in the 1840s, when Charles Babbage's *Analytical Engine* was proposed and Lady Ada Lovelace, Lord Byron's daughter, described how it could be programmed to compute Bernoulli numbers. While a visionary programmer and believer in Babbage's work, she nonetheless asserted: "the Analytical Engine has no pretensions whatever to originate anything. It can do whatever we know how to order it to perform" (Menabrea 1842, note G). This question also motivated Alan Turing's 1950 paper that defined the *Turing Test* for machine intelligence, which proposes that any machine that can fool you into thinking it is another human, is intelligent.

While intelligence is sought in conversation, creativity is usually looked for in art, poetry, and buildings. In each case, creativity combines a production system with an assessment filter (not all word-sequences are poetry). Production of a solution or expression, as we have seen, means defining a state in the problem's solution space. Creativity, on the other hand, is about *novelty*, *value*, and *surprise* or unexpectedness, elements of the assessment filter (Maher and Fisher 2012; Maher et al. 2013). A sizable literature has grown up around creativity assessment in the AI community, though most published research on computational design continues to rely on human judges.

Current research into AI divides the field into two parts. Expert chess players, diagnostic assistants for doctors, and a number of other so-called "narrow" AI applications form one part (Greenwald 2015), while "general" AI remains a goal just over the horizon. Narrow AI tools in the AEC industry may well facilitate aspects of design and construction practice such as optimizing façades or aspects of building operation. Research in these areas is occurring, but there is opportunity for more. A true design partner is probably an example of a general AI application, one that is still many years away (Gopnik 2016).

The Creative Designer

Fortunately, it may not be necessary to define creativity in order to support creative practice. We know that most design follows a pattern of creative or divergent thinking, followed by a convergent "working out" of the details, a pattern that may repeat with variations on a given project. While highly personal, designers often prepare for divergent thinking by doing comparison studies, diving into project details, brainstorming about alternative strategies or configurations, sketching (not editing), etc. There is opportunity for developing or tuning tools to facilitate activities such as image search, pastiche or collage, sketching, and collaboration through undemanding interfaces that support the state of flow, or immersive engagement.

Summary

Candidate solutions to design problems can be thought of as a set of parameters or values in the problem's solution space (its *genome* in a genetic space). In the absence of direct solutions or provably effective search strategies, research has focused on using computers to generate solution candidates systematically through rule-based mechanisms and parametric recombination (evolutionary algorithms), with candidates evaluated using a fitness function. Combined with scale- and phase-appropriate automated evaluation through simulation, statistics, or other techniques, these workflows present powerful ways of executing the convergent portion of the divergent–convergent design cycle.

Stronger support for exploratory divergent thinking, support which doesn't place much cognitive load on the user but which provides responsive assistance, is needed. These tools will probably use web-based image search, recognition, sorting, and tagging; and create media opportunities through mixed-modality sketching, collaging, and diagramming, all areas open to additional research.

Suggested Reading

Alexander, Christopher. 1964. *Notes on the synthesis of form.* Cambridge, MA: Harvard University Press.
Simon, Herbert. 1969. *The sciences of the artificial.* Cambridge, MA: MIT Press.
Suchman, Lucy. 1987. *Plans and situated actions: The problem of human–machine communication.* Cambridge: Cambridge University Press.
Woodbury, Robert and Andrew L. Burrow. 2006. Whither design space? *AIE EDAM: Artificial Intelligence for Engineering Design, Analysis, and Manufacturing* 20: 63–82.

References

Alexander, Christopher. 1964. *Notes on the synthesis of form.* Cambridge, MA: Harvard University Press.
Andia, Alfredo and Thomas Spiegelhalter. 2015. *Post-parametric automation in design and construction.* Norwood, MA: Artech House.
Cross, Nigel. 2011. *Design thinking: Understanding how designers think and work.* Oxford: Berg.
Dorst, Kees. 2004. The problem of design problems, in *Proceedings of design thinking research symposium* 6. http://research.it.uts.edu.au/creative/design/index.htm.
Dorst, Kees. 2011. The core of "design thinking" and its application. *Design Studies* 32: 521–532.
Gopnik, Alison. 2016. Can machines ever be as smart as three-year-olds? *Edge* (February 29). http://edge.org/response-detail/26084
Greenwald, Ted. 2015. Does artificial intelligence pose a threat? A panel of experts discusses the prospect of machines capable of autonomous reasoning. *Wall Street Journal (Online).* May 11.
Heisserman, J. 1991. *Generative geometric design and boundary solid grammars,* PhD Dissertation, Department of Architecture, Carnegie Mellon University.

Hornby, Gregory, Jason Lohn, and Derek Linden. 2011. Computer-automated evolution of an X-band antenna for NASA's space technology 5 mission. *Evolutionary Computation* 19: 1–23.

Lui, Ann Lok. 2015. Experimental logistics: Extra-architectural projects at SOM, 1933–1986. Unpublished Master's thesis, MIT.

Maher, M. L. and D. Fisher. 2012. Using AI to evaluate creative designs, in *Proceedings of the 2nd international conference on design creativity (ICDC2012)*. Edited by A. Duffy, Y. Nagai, and T. Taura, 45–54. Glasgow: Design Society.

Maher, M. L., K. Brady, and D. H. Fisher. 2013. Computational models of surprise in evaluating creative design, in *Proceedings of the fourth international conference on computational creativity 2013*. Edited by M. L. Maher, T. Veale, R. Saunders, and O. Bown, 147–151. Sydney: Faculty of Architecture, Design and Planning, University of Sydney.

Menabrea, L.F. *Sketch of the Analytical Engine invented by Charles Babbage*. Translated and with notes by Ada Augusta, Countess of Lovelace. From the *Bibliothèque Universelle de Genève*, October, 1842, No. 82. www.fourmilab.ch/babbage/contents.html.

Mitchell, Melanie. 1996. *An introduction to genetic algorithms*. Boston, MA: MIT Press.

Mitchell, William J. 1977. *Computer-aided architectural design*. New York, NY: Van Nostrand Reinhold.

Newell, Alan, J. C. Shaw, and H. A. Simon. 1958. Elements of a theory of human problem solving. *Psychological Review* 65: 151–166.

Rittel, Horst and Melvin M. Webber. 1973. Dilemmas in a general theory of planning. *Policy Sciences* 4: 155–169.

Simon, Herbert. 1969. *The sciences of the artificial*. Cambridge, MA: MIT Press.

Suchman, Lucy. 1987. *Plans and situated actions: The problem of human–machine communication*. Cambridge: Cambridge University Press.

Woodbury, Robert and Andrew L. Burrow. 2006. Whither design space? *AIE EDAM: Artificial Intelligence for Engineering Design, Analysis, and Manufacturing* 20: 63–82.

13
BUILDINGS

Computation Sources and Sinks

> [T]he integration of sophisticated sensing and actuation technologies for building-systems control will enable the benefits of understanding and responding to changing environmental conditions and the dynamic needs of the occupants.
>
> *Technology and the Future of Cities (PCAST 2016)*

> Using over 20,000 sensors, in The Edge it's possible to accurately monitor in real-time how the building is being used.... Everyone ... with a smartphone or tablet may ... regulate the light and the "climate" of his workplace.... [The lights] are only active when necessary ... intelligent technology can contribute to a smarter and more efficient use of buildings.
>
> *Edge developer OVG (2016)*

Every major technological shift has had an impact on the shape of buildings and cities, as well as the people that use them. The development of iron and steel enabled taller buildings, but without electricity to power and communicate between them, air conditioning to cool them, and elevators to make their heights accessible, we would still be building three- and four-story buildings such as we find in cities from the eighteenth and nineteenth centuries. Digital technology in the past 50 years has revolutionized both communication and computation. Today's smartphone users are texting or watching cat videos on a computer much more powerful than those that accompanied astronauts to the moon (Tomayko 1988). How have these changes altered our buildings, our cities, and our construction processes? How can architects integrate this technology into the design and production of new buildings in ways that improve building performance and occupant quality of life? How can government facilitate improved design, energy efficiency, and healthy, resilient communities through technological innovation? The two quotes above, one from a US White House report and the other describing a new building in the Netherlands, illustrate how deeply intertwined computing and the built environment are becoming.

In his book *Me++*, Bill Mitchell illustrated the impacts of digital technology on architecture by reviewing the architectural changes that banks have gone through in the past few decades (Mitchell 2003). At the end of the nineteenth century, banks needed to convey "trustworthiness" to customers so they would deposit their money with the bank. As a result, bank buildings were solid brick or stone edifices, with polished stone and wood-paneled interiors, as well as very visible and elaborate vaults. In the middle of the twentieth century the automobile inaugurated the drive-up bank, where you could do all your banking without stepping out of your car. In the late twentieth century, the anytime anywhere bank innovation—the ATM—meant a further reduction in bulk while delivering the same or greater functionality, perhaps in the local 24-hour convenience store. Today, electronic processing of deposits and payments is commonplace, supported by sophisticated computer systems, and we are even less concerned with the physicality of checks and cash. Money has been reduced to a balance displayed on the screen—a communications and computing issue. The banking application on your smartphone lets you check your balance, deposit checks, and move money around between accounts without entering a physical institution. As a result, some banks now do all their business online.

Some of the issues related to formal design of buildings and the social and legal structures that support design activity have been discussed in earlier chapters. In this chapter we'll have a quick review of the ways information technology is changing individual building use, as well as impacts on urban areas containing multiple buildings—what we hope will be the "smart cities" of the future.

Efficient Production

The White House report identifies four opportunities in the realm of buildings and housing: "(1) pre-fabrication, (2) modular construction, (3) customization and personalization, and (4) technologies for sensing and actuation" (PCAST 2016). The first two of these would mean replacing current practice, in which most construction is one-of-a-kind, both in terms of the construction team and the actual building configuration, with a more industrial approach. This is often cited as a factor in the "missing productivity" mentioned in Chapter 3, as is inclement weather conditions on exposed construction sites. Though often associated with cheap mobile homes and low-quality projects from the 1960s, these strategies are seeing more successful use today. Supported by strong BIM databases, project visualization, and customization opportunities (all tied to digital technology), off-site fabrication of panels or modules, possibly with the help of computer-controlled tools and robots, would allow experience to inform production, making it more efficient while reducing waste and risk to workers in a weather-proof environment, but without sacrificing quality or variation. The rise of modular design means buildings are increasingly being "installed" rather than constructed—in days rather than weeks or months (BDCNetwork 2012). Of course, neither all building designs nor all materials are suitable to this sort of construction; designers and fabricators need to work together closely to create or discover the best opportunities to deploy these strategies.

Beyond the Thermostat and the Water Cooler

Banks are not the only building type changed by technological developments. High-rise buildings use algorithms to dispatch elevators according to historical patterns of demand. They precool with night-time air when possible to reduce day-time air-conditioning loads. Businesses issue employees with ID cards that can be used to open parking garages, give elevator access to secure floors, or unlock doors. Some systems even monitor the location of employee cellular phones (via Bluetooth) to identify conference room use in real time, record meeting attendance, etc. (Robin 2016). Use of resources and building or room access can be tracked and monitored by managers, and scheduled or disabled depending on employee status. These changes to building operations impact system capacities used in designs, but others more directly impact building occupants. One new complex in Sweden goes so far as to offer tenants the option of inserting an RFID chip under the skin of their hand to provide access to doors and copiers (Cellan-Jones 2015).

One of the reasons we construct buildings is to bring information and people together in one place to work and communicate. To the extent and degree that digital technologies can deliver comparable experiences, we might reduce the need to build, but where will we spend our time—at home? Many young people today are both social and entrepreneurial. Their patterns trend more toward the short-term, limited-occupancy models of the sharing economy. New building-occupancy models and organizations leverage physical proximity to mix work and social life—the co-founder of WeWork calls its co-work spaces a "physical social network" (Rice 2015)—perhaps reducing the need for the non-work spaces, such as large suburban houses or freeways. Whether seen as a threatening or a promising development, it is an architectural concern, as real as cubicle-farms and corner-offices, lunchrooms, reception, and conference spaces. Understanding the affordances of architecture in the context of cultural change, business operations, and human behavior will help make digitally augmented offices or virtual environments more humane and productive.

Delivering a productive work environment to the individual is more than power, network, and a refrigerator or microwave oven in the common room. Individual needs, including thermal and lighting comfort, have been considerations in building design all along, but the advent of inexpensive wireless sensors and digital processing means the assessment and control of comfort can be more nuanced in space and time, and need not be aggregated around single-point sensors like thermostats, or single-point controls such as light switches or clocks. Building systems can often control lighting and mechanical systems almost to the level of individual desks. Lighting levels can be sensed and lighting adjusted automatically. Where day-lighting is a significant component of work lighting, blinds or shades are often adjusted by sensors today. In 2014 NBBJ developed and presented a prototype sunshade that could be adjusted at the level of individual glazing panels, based not on a schedule, but on exterior conditions, interior needs, and user wishes gathered through smartphone apps (Brownell 2014).

Managing Resource Use

As ongoing and substantial consumers of energy and water, the buildings and their operation have attracted attention for many years. Unfortunately, it is difficult to know if our actions have the intended effect. Measuring actual energy use at the building component level has required a great deal of expense installing wired sensor networks, so it is not done often. Measuring use at the building's electricity and gas meters can tell you overall energy consumption, but not much more. As a result we don't really know how well most building designs perform. New inexpensive sensors and ubiquitous wireless networks mean that much more can be done to validate simulation results and diagnose problem buildings. In addition, though such sensor networks traditionally required a substantial number of "point of service" sensors, recent research with single-point-of-contact sensors and machine-learning promises energy and water use monitoring with even greater cost reduction and greater data flows (Froehlich *et al.* 2011). Analyzing that data and drawing appropriate architectural, operational, and building-code lessons from it is likely to be a substantial area of research in coming years.

Building skins, long recognized as a domain of aesthetic expression by the designer, are also recognized as critically important components of the building's ability to deliver visual and thermal comfort to occupants without consuming vast amounts of energy. So-called "high performance" buildings that sip rather than gulp power are increasingly common and their control systems are almost always digitally monitored and directed.

At the same time, lighting is undergoing several revolutions. New construction often includes occupancy sensors that turn lights off when they think a room is unoccupied. New long-lived LED light sources work off of low-voltage circuits and may be individually addressable and dimmable, allowing the type and topology of wiring to change, while potentially eliminating the need for a wired wall switch. Mobile device apps and Wi-Fi networks deliver building status data to the desktop and take instructions, but do require network and power services.

Developing smart control strategies and systems such as those deployed in The Edge will be an important part of smart environments of the future. Research and development in this area of design computing will help to integrate such systems into routine design and operations so we use resources intelligently, and (hopefully) so we do not have to wave our arms periodically to prove we are alive to keep the lights on.

Our many electrical devices mean that wall outlets and their related "plug-loads" in offices now account for approximately 25 percent of building energy use (GSA 2012). While airports and university lecture halls were often built with only a few power outlets in the past, they are being remodeled to provide much more power for occupant laptops and associated devices, and building managers are looking for ways to distinguish between devices that are always on because they need to be (e.g., a computer server or electrical clock) and those that are always on because we don't turn them off (e.g., desktop printers), allowing power management systems to shut them down at night. As designers often determine or select the systems that interface between occupants and building systems, it makes sense for us to think about the best way for such interfaces to work and to be aware of available options.

Big Cities, Big Data, and the Effort to Get Smart

> It is posited that, through the integration of data services ... cities can be transformed.
>
> Boyle et al. (2013)

Beyond individual buildings, at the civic or urban design scale, the way we govern, work, and move about is changing. Many businesses adopt flexible hours and working arrangements, including telework from home. These policies ease commutes, but so do urban-scale data systems such as Seattle's real-time "e-park" off-street parking availability data stream (Seattle 2016). Smartphone apps that provide real-time scheduling or transit information (TFL 2016; OneBusAway 2016) effectively turn every smartphone or tablet into an information kiosk, just as they have already turned every coffee-shop into a field-office. The "Street Bump" project in Boston (New Urban Mechanics 2016; Simon 2014) enlisted phone-toting citizens across the city to help identify and locate potholes. Simply by running the Street Bump app on their accelerometer-equipped smartphone and traveling within the city, phones were turned into street-roughness sensors.

Obviously, while they may not require a wired connection to communicate, sensors still need power to run. Supplying that power creates a substantial installation expense that has limited our ability to take advantage of long-term sensing in buildings; sensing that might provide early warning of water infiltration, failures in door and window seals, rot, etc. Fortunately, continued miniaturization of electronics, research in peer-to-peer networks, and experience scavenging power from the sea of ambient energy radiated by normal power circuits in buildings is leading to the development of ultra-low-power sensor systems, sometimes referred to as "smart dust" (Hempstead et al. 2005). In outdoor settings, solar power is often available, such as solar-powered compacting garbage bins that help reduce resource consumption by eliminating unnecessary service trips (Bigbelly 2016). When both power and communication are wireless, many new opportunities to both gather and apply data become feasible.

The deployment of sizable embedded and walking sensor networks is allowing us to find efficiencies, but it may also allow us to check the assertions that have been made by restrictive-code advocates, engineers, and designers for years. Building owners and managers, as well as organizations such as the US Green Building Council, are accumulating information about the actual use and performance of their buildings, data that will demonstrate or debunk the efficacy of many conservation strategies. This is a relatively new phenomenon, but the potential for improved design implied by this feedback loop is substantial (Davis 2015). Its realization will require skilled and careful analysis, as well as political savvy.

Cities all over the world are trying to use widespread sensing to become more efficient—launching projects that aim to improve their governance and policing functions through digital technology. London is known for its use of closed-circuit TV (CCTV) cameras, but is just one of many cities that have instrumented city infrastructure and are making that information available. The city of Chicago, in collaboration with Argonne National Labs, is deploying sensors along city streets in an effort to quantify urban weather, air pollution, sound, and vibration, building an open-source real-time data stream. The goal is to study and ultimately understand what makes a city healthy, efficient, and safe (Sankaran et al. 2014). As our understanding of systems and processes in the urban setting comes into greater focus, it is likely that new design paradigms will emerge.

Virtuality and Buildings

As we leaf through a photo-album of summer vacation photos we revisit those places. The degree to which we disconnect from our physical surroundings and enter into the experience of the other reality is referred to as *immersion*. Different media, such as movies or stereoscopic image pairs, produce different degrees of immersion. *Virtual reality*, a term coined in the 1980s, refers to an immersive experience in which multiple sensory inputs, such as vision, hearing, and touch, are produced by digital technology, artificially producing a cognitive state of "being there" similar to that of the real world.

Virtual environments are often used for training and entertainment, and also hold promise for social-networking, shopping, and education experiences. These applications often utilize architectural or urban settings and metaphors. Though never intended to be instantiated in the real world, they work best when they "feel right"—when they are designed using the patterns and expectations of the natural world (Wonka et al. 2003). This represents an employment or research opportunity for designers, architectural historians, and other researchers who study the built environment.

Most virtual environments are created in traditional modeling software and only consumed in an immersive setting, a condition that most designers would like to change. As mentioned in Chapter 9, research is taking place regarding "in-world" interaction paradigms and instruments equivalent to mice or tablets that will enable a designer to edit the very environment they inhabit. Such tools, if accessible to homeowners, might revolutionize the field of design. They already challenge us to reflect on the core features of good design.

Other opportunities exist as well. Historical databases of typical designs, linked to vision-based analysis, might assist first-responders in active-shooter or natural disaster situations where limited real-time information is available. Augmented reality systems that overlay synthetic content on top of normal vision may give us the ability to place signage, advertising, and even entire building skins in the virtual realm while registering them with features of the physical world. These new technological opportunities all include elements of design and technology—the realm of the architect.

Summary

Increasing use is being made of digital sensing technologies in buildings, both during and after construction. These data streams can be monitored and compared to predictions to ascertain actual building performance, preemptively intervene to solve problems, and retroactively assess designer and contractor performance. Increasingly, developers, designers, and contractors are all looking to leverage this new data to make their operations more productive or their buildings safer, but each has unique data and processing needs. These developments are creating new opportunities for specialization and business. Ultimately, big-data research may help resolve questions that exist regarding disappointing building energy savings (Mehaffy and Salingaros 2013) in the face of claims of success (GSA 2008). The data will allow us to better connect our sophisticated numerical simulation tools to the real worlds of inhabitation, practice, and construction.

In the meantime, patterns of building use, from residential, to commercial office space and parking structures, are changing under the influence of communications and information technologies. Given the speed of those changes and the lifespan of most building projects, designers will need to be attentive to questions of building services flexibility, sensing, and intelligent response opportunities that enable environments to be custom fit to changing patterns.

Suggested Reading

Davis, Daniel. 2015. How big data is transforming architecture. *Architect Magazine*, April 23. www.architectmagazine.com/technology/how-big-data-is-transforming-architecture_o
PCAST. 2016. *Technology and the future of cities*. Executive Office of the President, President's Council of Advisors on Science and Technology, February.
Simon, Phil. 2014. Potholes and big data: Crowdsourcing our way to better government. *Wired Magazine*. www.wired.com/insights/2014/03/potholes-big-data-crowdsourcing-way-better-government

References

BDCNetwork. 2012. Modular construction delivers model for New York housing in record time. *Building Design + Construction*, September 19. www.bdcnetwork.com/modular-construction-delivers-model-new-york-housing-record-time

Bigbelly. 2016. Bigbelly waste and recycling stations. Bigbelly Solar Inc. http://bigbelly.com/solutions/stations

Boyle, David, David Yates, and Eric Yeatman. 2013. Urban sensor data streams: London 2013. *IEEE Internet Computing* 17: 12–20.

Brownell, Blaine. 2014. NBBJ develops a sunshade prototype: A dynamic, user-controlled canopy blocks infrared and sunlight. *Architect Magazine*. August 18. www.architectmagazine.com/technology/nbbj-develops-a-sunshade-prototype_o

Cellan-Jones, Rory. 2015. Office puts chips under staff's skin. *BBC News* www.bbc.com/news/technology-31042477

Davis, Daniel. 2015. How big data is transforming architecture. *Architect Magazine*, April 23. www.architectmagazine.com/technology/how-big-data-is-transforming-architecture_o

Froehlich, Jon, E. Larson, S. Gupta, G. Cohn, M. S. Reynolds, and S. N. Patel. 2011. Disaggregated end-use energy sensing for the smart grid. *IEEE Pervasive Computing, Special issue on smart energy systems* (January–March).

GSA Public Building Service. 2008. Assessing green building performance. www.gsa.gov/graphics/pbs/GSA_AssessGreen_white_paper.pdf

GSA Public Building Service. 2012. Findings, September 2012 plug load control study. www.gsa.gov/graphics/pbs/PlugLoadControl_508c.pdf

Hempstead, Mark, Nikhil Tripathi, Patrick Nauro, Gu-Yeon Wei, and David Brooks. 2005. An ultra low power system architecture for sensor network applications. *ISCA '05 proceedings of the 32nd annual international symposium on computer architecture*, 208–219. Washington, DC: IEEE Computer Society.

Mehaffy, Michael and Nikos Salingaros. 2013. Why green architecture hardly ever deserves the name. *ArchDaily*, July 3. www.archdaily.com/396263/why-green-architecture-hardly-ever-deserves-the-name.

Mitchell, William J. 2003. *Me++: The cyborg self and the networked city*. Cambridge, MA: MIT Press.

New Urban Mechanics. 2016. Street bump. http://newurbanmechanics.org/project/streetbump

OneBusAway. 2016. OneBusAway: The open source platform for real time transit info. http://onebusaway.org

OVG Real Estate. 2016. The Edge: OVG. http://ovgrealestate.com/project-development/the-edge

PCAST. 2016. *Technology and the future of cities*. Executive Office of the President, President's Council of Advisors on Science and Technology, February.

Rice, Andrew. 2015. Is this the office of the future or a $5 billion waste of space? *Bloomberg Business*, May 21. www.bloomberg.com/news/features/2015-05-21/wework-real-estate-empire-or-shared-office-space-for-a-new-era

Robin Powered, Inc. 2016. Meeting room booking system made easy: Robin. https://robinpowered.com

Sankaran, R., R. Jacob, P. Beckman, C. Catlett, and K. Keahey. 2014. Waggle: A framework for intelligent attentive sensing and actuation. *American Geophysical Union, Fall Meeting 2014*, abstract #H13G-1198.

Seattle. 2016. e-Park: Find parking faster in downtown Seattle. City of Seattle Department of Transportation. www.seattle.gov/transportation/epark

Simon, Phil. 2014. Potholes and big data: Crowdsourcing our way to better government. *Wired Magazine.* www.wired.com/insights/2014/03/potholes-big-data-crowdsourcing-way-better-government

TFL. 2016. Tube, overground, TFL rail & DLR status updates. City of London Transport for London. https://tfl.gov.uk

Tomayko, James E. 1988. Computers on board the Apollo spacecraft, in *Computers in spaceflight: The NASA experience.* NASA History Office. http://history.nasa.gov/computers/Compspace.html

Wonka, Peter, M. Wimmer, F. Sillion, and W. Ribarsky. 2003. Instant architecture. *ACM Transactions on Graphics* 22: 669–677.

14
PEDAGOGY
Teaching the New Designer

> To criticize a particular subject, therefore, a man must have been trained in that subject (1.3) ... not everybody can find the center of a circle, but only someone who knows geometry. (2.9) ... We learn an art or craft by doing the things that we shall have to do when we have learnt it (2.17).
>
> *Aristotle, Nicomachean Ethics*

The ancient Greek philosophers believed that education produces good or moral behavior. Aristotle divided education into two domains: reason and habit. The former involves understanding "the causes of things" while the latter derives from his belief that "anything that we have to learn to do we learn by the actual doing of it." This division continues to today. Students need to learn facts and theories, to have and organize experience (Bagley 1907), while repetition of certain actions helps convert them to muscle-memory, allowing higher-order thinking to guide their use. Donald Schön, one of architecture's most important educational theorists, emphasizes the role of the studio mentor as a live model for the "reflective practitioner" (Schön 1990) who demonstrates both execution skill and contextual understanding. This model recognizes that education may flow from either theoretical or practical wellsprings, reinforced by explanation and reflection in the case of theory, and by repeated practice in the case of the practical. In the process of teaching someone, we equip him or her with conceptual and sometimes physical tools and skills with which they understand the world and address its problems.

Traditional design curricula, and architecture in particular, have emphasized a well-rounded education balancing practical knowledge and theoretical understanding in a number of subjects, skill with representational media such as drawing, painting, and sculpture, but dominated by the repeated practice of design problem solving.

Unfortunately, these curricula often view digital technology as simply another medium in which to complete traditional work within the academic program and to secure employment. There is often little room for the topics discussed in this book, nor for development of the habits, skills, and knowledge needed to advance their study—the study of design computing. As things stand, developing robust curricula prototypes is one of the field's challenges, though implementation is probably a larger obstacle.

As far back as 1969 economist and Nobel laureate Herbert Simon proposed a "Science of Design" that would complement the "Natural Sciences" of physics, chemistry, biology, and so on (Simon 1969). Such a discipline would bring a scientific methodology to many of the questions this book has sought to tease out, including understanding the challenge of problems and the cognitive needs of design, data structures for representing design problems and artifacts, problem definition and search of design-space, generation and refinement of candidate solutions, plus synergy and leverage opportunities to combine human and machine intelligence.

Unfortunately, while mastery of representational media is essential to both pedagogy and practice of architecture, digital media are rarely embraced with enthusiasm. While most faculty members have historically had some level of skill with traditional media, digital media has been the realm of expertise. As a result, schools have tended to separate and isolate digital media, often focusing solely on skills in the *operation* of particular software packages, and offering the material in *auxiliary* digital media courses. This has reinforced a tendency to define the subject in terms of production, a task-oriented approach to design computing that equates CAD with AutoCAD, 3D modeling with Rhino or 3DS Max, and BIM with Revit. In particular, this approach has not challenged a common conceit from the 1980s that CAD is "just another pencil," that it doesn't matter, because digital tools can be exchanged on a one-for-one basis with traditional practices and tools, and that such swaps are "neutral" with regard to individual cognition or broader professional impact. I recall a faculty meeting in the mid-1990s, during which a colleague asserted that we did not have to provide computer-literacy instruction at all, because the next-generation student would bring the relevant knowledge with them. How this argument could be made in a profession that routinely teaches grown-ups how to use a pencil is a mystery, but it is consistent with the anxiety-reducing notions of one-to-one substitution and tool-neutrality.

However, when new tools are fundamentally different from old tools, new behaviors and new ways of thinking about problems tend to emerge. In the past, constructing a perspective was laborious, so students tended to stick to plan and elevation views, with occasional axonometric drawings. When they did construct a perspective, care was taken in finding an appropriate viewpoint in order to compose a powerful image. Modern 3D rendering software makes *production* of realistic perspectives straightforward, but it doesn't help *select* revealing viewpoints, and sophisticated rendering can give both clients and students the impression that a design is further along than it really is. Aristotle would have recognized that both situational insight and operational skill are required to make best use of the new tools.

Digital Natives vs. Digital Immigrants

One prominent exception to the neutral-tool view that has had significant press exposure is found in two publications from 2001 in which Marc Prensky developed a distinction between "digital natives" (meaning those born after 1980 or so) and "digital immigrants" (those born earlier), with particular attention to education (Prensky 2001a, 2001b). To paraphrase: The digital natives have grown up with computers and personal access to computing and multi-media. Their manipulation of the technology is said to be missing the "accent" (clumsiness?) of the immigrants, but (alas!) the immigrants are in charge of the classroom. Overall, the articles argue for rethinking how material is presented to "digital natives" in favor of hyperlinked multi-tasking multimedia games. That the "Myth of Multitasking" has been largely debunked (Rosen 2008) in the subsequent decade has not stopped the natives vs. immigrants dichotomy from entering common usage.

Oddly, students often seem attracted to this theory. One reason might be that the theory justifies their attraction to cellphones and laptops, and relieves them of responsibility for engaging the challenging changes wrought by digital technology and adjusting their own behaviors in response—they are, after all, unable to escape their own accent.

Implicit in the dichotomy is a kind of manifest destiny—a belief that natives are more competent users than their predecessors. This certainly resonates with the broader cultural adulation of technology consumption in general and youth in particular, and my observations above about digital media courses. My own observation is that modern students are much more comfortable with technology than are their predecessors, which is very good, but that their education in the fundamental principles of the technology has been ad hoc and haphazard, focused on task-completion rather than deep understanding. As a result, most students are unafraid of the technology and familiar with the first-level affordances, but may not be comfortable with more subtle subjects and don't completely "own" their technology.

The debate sparked by Prensky's articles continues, as does related discussion and research about whether the multi-tasking enabled by modern technology is a good thing or a bad one, whether premature materialization made possible in modern 3D visualization software seduces students into thinking they have done good design, etc. It is perhaps telling that Prensky wrote a follow-up piece that focused on digital *wisdom* rather than skill, on reason rather than habit (Prensky 2009).

The Collapsing Options Model

Traditionally, architectural education has focused on questions of design, set against a backdrop of technical assessment and augmented by media manipulation skills offered in what are all-too-often called "support" courses. As noted above, with the exception of a few schools where "digital" or "paperless" studio experiments were undertaken, the advent of digital analysis and media options has been addressed with specialized course content focused on acquiring skills with the new tools, each one an option on top of the traditional structure. As more and more of these courses have been launched, covering more and more of the content material from the traditional curriculum, they function less as specialized advanced-skills conduits and more as entries to the mainstream of the profession. Unfortunately, because of the elective and ad-hoc nature of this coursework, appropriate skills and expectations are rarely integrated across mainstream course and studio content in an appropriate fashion, often because the instructors of these less-technical courses lack paradigms for such instruction in their own backgrounds, being, themselves, digital immigrants. The result of these entrenched interests (both within universities and their accreditation bodies) means that many students are ill-prepared to work with the full range of tools available in the workplace, and their exposure to them is haphazard.

As a profession, architecture in the Western world has not expected to graduate finished architects directly from school; there has always been an expectation of experience and polish to be developed during internship, working under the direction of senior designers. As discussed in Chapter 10, however, these relationships have changed as firms seek new knowledge and skills through hiring. Who is best prepared to teach this material? When is the best time to learn it? How do we get beneath the surface froth of change to the deeper currents of transformation? These are some of the questions that beset teachers and stewards of curricula in design and computing. Waiting for the previous generation to retire will not be enough; we need good models that address real design challenges with appropriate digital media skills in a reflective and considered fashion that takes advantage of the technology's affordances and avoids the pitfalls. In their absence, other models will take hold, models that will likely continue the mercenary task-oriented view of the technology.

Where is the knowledge of design computing best located in a design school curriculum? Perhaps the changes are so ubiquitous that traditional media should be abandoned and digital media use should be integrated across the entire curriculum? Such an approach would require that faculty have motivation, time, and opportunity to rethink and restructure their own teaching. Nonetheless, teaching the teachers might make the most sense. Such an approach would help with another problem as well—reliance on the elective course model means few students can take and few curricula can offer well-organized subject-area *sequences* in design computing within a professional degree program without becoming overlong.

The curriculum is a reflection of the faculty, and managing the faculty resource is a challenge for department administrators. In the pre-digital era, most teachers of design had, like their practitioner cousins, substantial skill with traditional media, and were not hesitant to guide students in their representational choices. Presently there are many different software packages available, but a fairly small population of instructors with broad skills in the use of software, and very few who have mastered more than a handful. It is difficult for instructors to maintain broad mastery when speed and arcane knowledge are most valued, and it is difficult to staff a generalized presentation of the material.

It is important that we try to persuade students to take a longer view, to understand that the tools they are learning to use now will be replaced several times during their careers, that there is long-term value to focusing on more generalized knowledge in addition to mastery of the current tools. Aristotle's conclusion that damage occurs as often from too much as from too little of something is relevant here—finding a balance point is important.

What to Teach?

In response to the complexity of the industry and the limitations of tight curricula, schools often adopt simplifying policies, selecting a limited palette of software for use in their courses. While effective in reducing the learning curve that students must climb, and saving money on software licenses, this strategy renders some opportunities inaccessible, reinforces the one-size-fits-all task-oriented tunnel-vision of students at the expense of deeper understanding, and incurs risks associated with selecting or sticking with software that fails to enter, or falls out of, the mainstream of professional practice.

Lifetime Learning

A university-level design computing course needs to offer knowledge with a longer shelf-life than most software study guides, counteracting the cultural drift of the ad-hoc informal learning environment while injecting deeper knowledge into a task-oriented culture (Senske 2011). A software technology such as BIM is built on foundations of data-management, representation, interface affordances, and cultural integration. Including such general knowledge in a "BIM class" would be appropriate because it informs questions of data exchange, technology trends, and possible add-ons, but touching on the many constituent parts in other courses would begin building the foundation and enhance understanding.

Students *do* have legitimate task-oriented design computing skill needs, as well as interests that lie outside of design computing. It is fortunate that they are comfortable seeking out and using learning resources wherever they find them, because they may not have time to study the subject more deeply. In schools where students spend many hours in the studio doing tasks together, this means they often learn these skills either from one another or from the web. Unfortunately, it is difficult for this learning to get beyond surface knowledge and second- or third-hand knowledge often becomes corrupted or distorted in plausible but misleading ways.

Self-directed learning should be an element of life-long learning, a pattern that should include setting aside time to learn and expecting to share insights or skills with others. Small student-run workshops, faculty-led short course modules, term-long exercise sequences, and explicitly theoretical reflective seminars that address foundational questions—these are each important components of a habit of continuous learning.

Algorithmic Thinking

Learning to design involves learning a vocabulary. Traditional design dealt in symmetry, proportion, light, history, and material. Digital design adds to or displaces these with concepts rooted or transformed by the vocabulary of computers—sometimes referred to as algorithmic or computational thinking. The most obvious way to learn this vocabulary is through the experience and craft of scripting—generating form or image using computer code (Senske 2011). With the advent of visual programming languages, scripting is seen as much more accessible and interesting than that found in text-based environments, a change that significantly broadens its appeal.

In the 1980s the only way to study computing in architecture was to learn to code, whether in a language like FORTRAN or an embedded language like AutoLISP. Code was the primary way to explore design computing ideas. There was an explosion of code, some of which has coalesced into current end-user tools, allowing the next generation to adopt the task-orientation mentioned above. The current popularity of parametric design reflects a similar interest in personal control and expression, but it also responds to the widespread appearance of application programming interfaces (APIs) in software, and resonates with a national portrayal of code as a driver of innovation and economic development.

Algorithms have been around far longer than computers. Any pattern or sequence of actions that is designed to produce a predictable result can be thought of as an algorithm. Weaving patterns on looms, cooking, cutting or bending patterns in metal and wood, brick-laying, parquetry—all of these utilize repeated patterns of action. Designers often acknowledge thinking systematically, working from some set of principles or rules to a complete design. What is different today is the potential for the patterns themselves to change across the body of work, something that wasn't commonly done in the past. Where Renaissance architects might create intricate nested arrangements of proportion, today's designer may well design a building skin where every panel responds to the unique conditions of its orientation and solar exposure in a specific way, as on the façade of the London City Hall. This kind of thinking calls on the mind as well as the eye, a kind of thinking that has not been central to architecture for some time. It consists of variable geometry and relationships in a dance with fabrication and erection. Learning to create, think about, and control these complex structures is important in the modern world.

Considering Side-Effects

There are deeper questions in the area of pedagogy than whether to teach coding or require a course in BIM. Once we understand that digital tools are not simple one-for-one substitutes for their antecedent tools, we have to ask about the cognitive and social side-effects of using these new tools. In reviewing the cognitive aspects of design, the link between sketching and design has already been discussed.

What new self-delusion traps does the technology create? What new work disciplines are required besides "save often!" and "make backups"? What is the impact on the social community of the studio? Is digital work different from paper work? What pedagogic guidelines should we adhere to?

The affordances of a modern GUI are seductive. Nigel Cross, a noted design researcher in the UK, has observed: "Some student designers can seem to get stuck in gathering information, as a kind of substitute for actually doing any design work" (Cross 2011). The rendering speed and ease of manipulation available in modeling software today makes it easy to postpone the difficult business of design in favor of spinning the model a bit for a slightly different view, zooming in for a look over there, and so on. Even at schools that traditionally privilege technology, such as Carnegie Mellon University, in subjects that hinge on technology, such as interaction design, an interest in developing communication and thinking skills has led at least one instructor to a "pencils before pixels" approach (Baskinger 2008).

Observations of students using 3D modeling programs in design raise similar questions—after encountering a problem in design development, students are often seen to change their focus to another part of the model, or simply spin the model this way and that in what appears to be a random fashion, rather than isolate the problem and focus on resolving it. In the face of uncertainty regarding appropriate action, it seems possible that the affordances, or *opportunities* for action inherent in the interface, are so compelling that action trumps contemplation.

On top of this, the imagery produced by today's software is remarkable—textures, light sources, diffuse and specular reflections—all readily available whether or not the underlying design is good. It is easy to see how someone could confuse high-quality rendering with high-quality design. In his recent book, *The Death of Drawing*, architect and academic David Scheer (2014) criticizes the limited, self-defined world of (digital) "simulation" in contrast to the conceptually engaging realm of "representation." Design education involves learning to critique your own work, to look deeply at a project, learning to see not only in visible light but at the "abstract" end of the spectrum. This requires context.

Though we describe design as a cyclic or iterative process, a description that paper drawings supported through tracing and re-drawing (at different scales), our digital tools do not. The coordinates in an initial sketch, perhaps drawn at a very small scale on the screen, are indistinguishable from those of a carefully reviewed plan. What role does tracing play in developing a design? Redrawing, though inefficient in terms of drawing production, directs focused attention to selected areas of the project, offering a chance to recapitulate design rationale and reinterpret existing or emergent conditions. What, if anything, has taken its place?

While David Scheer's book confronts the possibility that overreliance on digital tools and simulation is diminishing architecture conceptually, there is another disturbing possibility linked to the modern reliance on digital tools. There is research with children that links hand-lettering and development of writing skill (Richards *et al.* 2009). This raises the possibility that sketching by hand is actually critical to learning to function as a designer. If so, it would explain the observation by James Cutler, mentioned in Chapter 7, that recently hired interns in his office seemed less able than their pre-digital peers to develop design ideas from sketches. Further, Cutler said that principals in firms located all across the United States had noted the same decline. It is, of course, very hard to verify such an observation, as each generation of observers is different, but if it is true, then schools of architecture should be concerned regardless of the cause. Cutler attributed the decline to under-involvement in sketching. If the cognitive decline is real, it might equally well be a question of how the discipline of creating or using digital representation is dealt with in the studio, not as a question of simple technical prowess, nor as a buzz-term to enhance reputation, but as a serious question engaged by studio mentors and students alike. The theory of *desirable difficulty* has been used to explain how students taking lecture notes using pen and paper learn better than students using a laptop (Mueller and Oppenheimer 2014). Perhaps something similar happens with sketching.

It is time to jettison the comforting notion that CAD is "just another pencil" once and for all, and develop evidence-based approaches to teaching design that fully acknowledge the changes that are accompanying the adoption of digital tools.

Summary

Digital technology has disrupted much in design, construction, and inhabitation of the built environment. Approaching design computing as simply a means of document production is shortsighted and simple-minded. These tools change what we can make and also seduce us into new mistakes. Developing a longer-range view that provides students with judgment, heterogeneous skills, seed knowledge, and habits to support a lifetime of ongoing disruption and change should be our goal. This is not a question of what a design computing curriculum should be; rather, it is a question of what a design curriculum should be.

Suggested Reading

Mueller, Pam A. and Daniel M. Oppenheimer. 2014. The pen is mightier than the keyboard: Advantages of longhand over laptop note taking. *Psychological Science* 25: 1159–1168.
Schön, Donald. 1990. *Educating the reflective practitioner: Toward a new design for teaching and learning in the professions.* San Francisco, CA: Jossey-Bass.
Senske, Nicholas. 2011. A curriculum for integrating computational thinking, in *Proceedings of ACADIA regional conference 2011*, 91–98. ACADIA.
Simon, Herbert. 1969. *The sciences of the artificial.* Cambridge, MA: MIT Press.

References

Aristotle. *Nicomachean ethics*, from *Aristotle in 23 volumes, Vol. 19*. Translated by H. Rackham. 1934. London: William Heinemann Ltd. www.perseus.tufts.edu/hopper
Bagley, William. 1907. *The educative process.* London: Macmillian & Co.

Baskinger, Mark. 2008. Pencils before pixels: A primer in hand-generated sketching. *Interactions* 15: 28–36.

Cross, Nigel. 2011. *Design thinking: Understanding how designers think and work.* Oxford: Berg.

Mueller, Pam A. and Daniel M. Oppenheimer. 2014. The pen is mightier than the keyboard: Advantages of longhand over laptop note taking. *Psychological Science* 25: 1159–1168.

Prensky, Marc. 2001a. Digital natives, digital immigrants. *On the Horizon* 9 (5).

Prensky, Marc. 2001b. Digital natives, digital immigrants, part 2: Do they really think differently? *On the Horizon* 9 (6).

Prensky, Marc. 2009. H. sapiens digital: From digital immigrants and digital natives to digital wisdom. *Innovate* 5 (3).

Richards, Todd, V. Berninger, P. Stock, L. Altemeier, P. Trivedi, and K. Maravilla. 2009. fMRI sequential-finger movement activation differentiating good and poor writers. *Journal of Clinical and Experimental Neuropsychology* 29: 1–17.

Rosen, Christine. 2008. The myth of multitasking. *The New Atlantis* 20: 105–110.

Scheer, David Ross. 2014. *The death of drawing: Architecture in the age of simulation.* Oxford: Routledge.

Schön, Donald. 1990. *Educating the reflective practitioner: Toward a new design for teaching and learning in the professions.* San Francisco, CA: Jossey-Bass.

Senske, Nicholas. 2011. A curriculum for integrating computational thinking, in *Proceedings of ACADIA regional conference 2011*, 91–98. ACADIA.

Simon, Herbert. 1969. *The sciences of the artificial.* Cambridge, MA: MIT Press.

INDEX

Page numbers in *italics* denote figures.

2D graphics 52–7, 68; drafting software 11, 54–7, 58, 113, 129–31; raster graphics 53–4, *56*, 57, 113, 118; sketching software 53–4, 57, 67–8, 113, 118; storing drawing data 55–7, *56*; vector graphics 53, 54–7, *56*, 58, 113, 115

3D graphics 57–67, 68–9; boundary representation (b-rep) models 59–61, *59*, *60*, 63–4, *64*; constructive solid geometry (CSG) models 61–2; conversion between representations 63–4, *64*; graphics pipeline 57, 64–7, *65*; non-manifold geometry 62, *63*; photorealistic images 67, 146; point clouds 58, 62; solid models *59*, 61–2, 63–4, *64*; voxels 58, *59*, 62; wireframe models 58–9, *59*, 63–4, *64*

3D printing 115, 132

addition 107
adjacency graphs 116
AEC *see* architecture, engineering, and construction (AEC) industry
affordances 11, 110, 131, 190
AI (artificial intelligence) 10, 125, 163, 168
Akin, Ömer 84
Alexander, Christopher 3, 78, 90, 94, 164

algorithmic thinking 189
algorithms 8–9, *9*, 13, 35, 57, 107–8, 189; genetic 70, 165, 168–9; machine-learning 168, 178; shading 60, 66; smoothing 60, 61, 66
alternative programming environments 44
ambiguity 96; in design documents 31, 143
American Institute of Architects (AIA) 54
American Standard Code for Information Interchange (ASCII) 37, 47–8
analogue models 21
analogy 90, 164, 170
animations 52, 58
application programming interfaces (APIs) 189
AR *see* augmented reality (AR)
Archea, John 150
architecture, engineering, and construction (AEC) industry 25–32; data 30–2; missing productivity gains 27–9, *28*, 141–2, 176
Argonne National Labs 180
Aristotle 184, 185, 188
artificial intelligence (AI) 10, 125, 163, 168
ASCII 37, 47–8
assignment statements 40–1
associative thinking 87

attributes 55, 108
Augenbroe, Godfried 32, 156, 157
augmented reality (AR) 132, 135
Autodesk Drawing Exchange Format (DXF) 56
AutoLISP 119, 189
automated analysis tools 70

Babbage, Charles 172
backtracking 88, 124–5
banks 176
BAS *see* building automation systems (BAS)
Beeman, Mark 84, 85
big data 149–50, 168
BIM *see* building information modeling (BIM) software
binary numbers 37, *37*, 45
bits 36–8, *37*
blocks/symbols 55
Boeing Company 160
Boolean data 39
Boolean operations 61
booting 36
BOP *see* Building Optimization Program (BOP)
Boston 179
boundary polygons 107–8
boundary representation (b-rep) models 59–61, *59*, *60*, 63–4, *64*
Boyle, David 179
bubble diagrams 62, *63*, 115, 116
building automation systems (BAS) 71
building codes 31
building information modeling (BIM) 69–70
building information modeling (BIM) software 23, 39, 110, 141, 161; and building operations 71, 147; in construction 143; government sponsorship of 28–9, 81, 141, 146–7; MacLeamy Curve 139, *140*
building operations 71, 147, 177–9
Building Optimization Program (BOP) 171
buildingSMART 32
built environment 3, 4–5, 25–32, 175–81; cities 179–80; data 30–2; efficient production 176–7; Internet of Things

(IoT) 5, 29–30; missing productivity gains 27–9, *28*, 141–2, 176; resource management 178–9; technology in buildings 5, 29–30, 177–9; and virtual reality 180–1
bytes 37, *37*, 45

CAD systems 11, 23, 68–9, 119, 122, 161; *see also* 2D graphics; 3D graphics
carbon dioxide emissions 4
Carnegie Mellon University 168, 190
Carr, Nicholas 95
carriage return (CR) 48–9
carriage return line-feed (CRLF) 49
case-based reasoning (CBR) 93, 119
central processing units (CPUs) 8, 35
certainty 95–6
characters *37*, 39, 47–8
Chicago 180
circulation, representation of 116–17, *116*
clash detection 32, 69, 143
classes 40
Clayton, Mark 5
climate change 4
clipping process 65–6, *65*
Close, Chuck 84, 85
closure 111–12
cloud-computing 67
CMC *see* computer mediated communication (CMC)
CNC *see* computer numerically controlled (CNC) machines
codecs 53
code points 48
coding *see* computer programming; scripting
co-evolution of problem and solutions 80–1, 87, 97, 98
collective memories 93
color depth 53
color raster displays 58
complexity 13, 112
compression/decompression schemes 53
computer aided design (CAD) *see* CAD systems
computer mediated communication (CMC) 136

computer numerically controlled (CNC) machines 70–1, 127
computer programming 35, 39–44; algorithmic thinking 189; alternative programming environments 44; assigning values 40–1; conditional execution 42–3; constants 40; data types 39–40; function declarations 41–2; functions 41–2; if-then statements 42–3; loops 43; parameters 41; parametric design 70, 71, 119, 189; pseudo-code 40; repetitive execution 43; scripting 9, 39, 70, 110, 119, 189; variables 40; *see also* standard representations
computers/computing 3–4, *3*, 8–10, 11–12, 34–8; affordances 11, 110, 131, 190; bits 36–8, *37*; bytes 37, *37*, 45; computational complexity 13; files and directories 49–50; graphics processing units (GPUs) 66; Internet of Things (IoT) 5, 29–30; memory 8, 35, 36, 38; operating systems 36; representation 8–9, *9*; secondary storage 35–6; sequentiality 38; Turing machine 10; virtuality 10, 35–6; viruses/malware 38; *see also* algorithms; interfaces; software/tools
computer supported collaborative work (CSCW) 136
conditional execution 42–3
connectivity graphs 116, *116*, 117
constants 40
constraint-based modeling 39
constraints 80, 96–7
construction: fabrication systems 14, 70–1, 91, 119–20, 127, 129, 142, 143, 148–9; missing productivity gains 27–9, *28*, 141–2, 176; modular 176–7; process changes 32, 142–3
Construction Research Congress 31
Construction Users Round Table (CURT) 141
constructive solid geometry (CSG) models 61–2
continuous learning 151, 188
contractual risk and reward structures 29

convergent thinking 87, 123
coordinate systems 46
copyright, fonts 49
cosine shading algorithms 66
cost functions *see* objective functions
CPUs (central processing units) 8, 35
creativity 86, 171–2
Cross, Nigel 84, 162, 163, 190
CSCW *see* computer supported collaborative work (CSCW)
CSG models 61–2
customization 176; mass 14, 119–20, 129
Cutler, James 99, 191
cyber tourism 92

data 12, 30–2, 35, 39–40, 122, 143–6; file-format obsolescence 148; Industry Foundation Classes (IFC) 12, 40, 122, 144; knowledge loss 144, *145*; reuse of 143–4; *see also* standard representations
data exchange standards 40, 122
data models 34
Davis, Daniel 147
Day, Martyn 69
day-lighting 178
Denver Art Museum 142
design *3*, 5–8; *see also* design process
design–bid–build model 29, 141
design–build–operate (DBO) models 29, 160
design cognition 84–100; case-based reasoning (CBR) 93, 119; certainty, ambiguity, emergence, and flow 95–7; collective memories 93; creativity 86; and decline of hand drawing 99–100, 190–1; design as selection 91, 144; design as social action 97–8; design by analogy 90, 164; design moves 88; and design process 86–9, *87*; desirable difficulty theory 100, 110, 191; developing memories 92–3; divergent and convergent thinking 87, 123; and drawing 88–9; issue-based information systems (IBIS) 93, 118–19; memory-based design process 90–5; pattern languages 94; problem solving through

subdivision 91–2, 162; and representation 106–10; and scale 89, 99; shape grammars 95, 118, 119, 168; side-effects of digital tool use 99–100, 189–91; and situated action 97–8; space syntax 94; and tools 98–100
design methods 4, 97, 163
design moves 88, 168–9
design problems 77–82; constraints 80; defining 78–9; objective functions 14, 80, 164, 166; performance goals 80; puzzle making 80–1, 87, 97, 98; situated action 78, 97–8, 164; state spaces 1, 12, 79–80, 163, 164–5, 167–9; stopping rules 14, 81–2, 171; subdivision of 91–2, 162; wicked problems 78, 97, 98, 164; *see also* solutions
design process 6–8, 7, 79, 86–9, 87; design moves 88; memory-based 90–5; role of drawing in 88–9
design rationale 118–19
design spaces 1, 12, 79–80, 163, 164–5, 167–9
desirable difficulty theory 100, 110, 191
digital fabrication systems 14, 70–1, 91, 119–20, 127, 129, 142, 143, 148–9
digital natives vs. digital immigrants 186
digital wisdom 186
direct manipulation 128, 130
directories 49–50
disembodied-participant problems, in virtual reality 135
divergent thinking 87, 123
divide-and-conquer approach 91–2, 162
Dorst, Kees 90
Dourish, Paul 127, 128
drafting software 11, 54–7, 58, 113, 129–31
drawing 88–9, 105, 112–13; drafting software 11, 54–7, 58, 113, 129–31; drawing coordination problem 112; effects of decline of hand drawing 99–100, 190–1; expert interpretation of 157, 160–1; incompleteness of drawings 157; meaning in 131; sketching software 53–4, 57, 67–8, 113, 118; storing data 55–7, *56*; symbolic content in 113–15, *114*
Dynamo 119

Ecotect 31
education 184–91; algorithmic thinking 189; desirable difficulty theory 100, 110, 191; digital natives vs. digital immigrants 186; elective course model 187; life-long learning 151, 188; and side-effects of digital tool use 99–100, 189–91; and task-oriented approach 34, 185, 186, 188, 189
Eisenman, Peter 61
embedded knowledge 108–10, *109*
emergence 96, *96*, 117–18
end-of-line 48–9
endpoint snap 55, 130
energy efficiency 4
Energy Information Administration, US 4
energy simulation 149–50
energy use monitoring 178
Engelbart, Douglas 127, 128
environmental viewing models 65
error sources 22–3
ethics 140
Euler's Formula 61, 63
expertise 154–61; building in-house 159; framing problem 157; interpretation of drawings 157, 160–1; leaf-node knowledge 159–60; levels of 90; round-trip problem 157–9, *158*
exponential notation 45
eye-space 65

fabrication systems 14, 70–1, 91, 119–20, 127, 129, 142, 143, 148–9
facilities management (FM) 71, 147, 177–9
Falling Water 89
Fenves, S. J. 32
files 49–50; duplication of 23; extensions 50, 122; file-format obsolescence 148; formats 50, 122
finite element models 20–1, 22
fitness functions 14, 80, 164, 166

flat shading algorithms 66
floating point numbers 39, 45–7
floor plans 62, *63*
flow 97, 132
fonts 49
FORTRAN 189
framing problem 157
Franken, Bernhard 149
Frederic C. Hamilton Building 142
function declarations 41–2
functions 41–2

GA *see* genetic algorithms (GA)
Gehry, Frank 61, 148
General Services Administration (GSA), US 28, 81, 141, 146, 147
Generative Components 119
generative rules 95
genetic algorithms (GA) 70, 165, 168–9
geographic information systems (GIS) 94
Georgia Institute of Technology 94
Gero, John 79, 84, 88, 96
gesture interfaces 133
Gibson, J.J. 11, 131
Gips, James 95
GIS 94
GKS metafile projects 57
glyphs 49
GMW Architects 150
Gouraud shading 66
GPUs *see* graphics processing units (GPUs)
grammar rules 26
grammars, shape 95, 118, 119, 168
graphical user interface (GUI) 11, 190
graphic attributes 55, 108
graphics 11, 51, 52; *see also* 2D graphics; 3D graphics
graphics pipeline 57, 64–7, *65*
graphics processing units (GPUs) 66
graphs 115, 116–17, *116*
Grasshopper 119
Graves, Michael 139
Green Building Council, US 180
green building design 149–50
Guggenheim Museum, Bilbao 61, 148
GUI *see* graphical user interface (GUI)

HCI *see* human–computer interaction (HCI)
Hellmuth, Obata & Kassabaum 150
Hem-Fir 19–20
hidden-line images 59
hidden-line removal 58, 66, 123
hidden-surface removal 58, 66
hierarchy of human needs 25
high performance buildings 178
hill climbing strategy 165, 167–8
HOK 150
Holland, John 168–9
Holocaust Memorial, Berlin 61
HTML 49
human–computer interaction (HCI) 78, 128–9
Hunt, Earl 90

IBIS *see* issue-based information systems (IBIS)
IFC *see* Industry Foundation Classes (IFC)
if–then statements 42–3
IGES *see* Initial Graphic Exchange Standard (IGES)
immersion 180
immersive technologies 133
Industry Foundation Classes (IFC) 12, 40, 122, 144
inference engines 130
infinity 38
in-house research groups 150
Initial Graphic Exchange Standard (IGES) 12
integers *37*, 39, 45, 46–7
integrated digital fabrication workflows 142
integrated project delivery (IPD) 29, 69, 160
intellectual property: BIM software 70; building codes 31; fonts 49; representations 45
interfaces 9, 11, 110, 127–37; affordances 11, 110, 131, 190; appropriate 132–3; augmented reality 132, 135; computer mediated communication (CMC) 136; computer supported collaborative work

(CSCW) 136; deciphering intent 129–31; direct manipulation 128, 130; gesture and voice 133; graphical user interface (GUI) 11, 190; human–computer interaction (HCI) 78, 128–9; inference engines 130; interactive paradigm 127, 128–9; opportunities to improve 131–6; snaps 55, 57, 130; ubiquitous computing 135–6; virtual reality 133, 134–5
International Alliance for Interoperability (IAI) 32
Internet of Things (IoT) 5, 29–30
internship 187
interoperability 12, 31–2, 144
intersection computation 112
IoT *see* Internet of Things (IoT)
IPD *see* integrated project delivery (IPD)
isovists 117
issue-based information systems (IBIS) 93, 118–19

Joint Computer Conference (1968) 127

Karlen, Ingvar 32
knowledge, embedded 108–10, *109*
Kounios, J. 84, 85
Kuhn, Thomas 92

Lasdun, Denys 162, 168
Lawson, Bryan 84, 104, 125
leaf-node knowledge 159–60
Le Corbusier 25
Libeskind, Daniel 142
life-long learning 151, 188
lighting control 177–8
line-feed (LF) 49
line-of-intersection computations 112
Logo programming environment 44
London 180
London City Hall 189
long-term memory 85, 123
loops 43
lossless codec 53
lossy codec 53
Lovelace, Lady Ada 2, 172
Lui, Ann Lok 171

McDermott, Terry 148
machine-learning algorithms 168, 178
MacLeamy, Patrick 139
MacLeamy Curve 139, *140*, 160
macros 44
mainframe computing 47, 51, 58, 69
malware 38
Mars Climate satellite 46
Maslow, A. H. 25
mass customization 14, 119–20, 129
MAXScript 119
memory, computer 8, 35, 36, 38
memory, human 85, 106; collective memories 93; developing memories 92–3; long-term 85, 123; memory-based design process 90–5; working 85, 106
microcomputers 51, 58
Mitchell, William 96, 98, 139, 176
model-in-model approach 134
models 19–23; analogue 21; finite element 20–1, 22; sources of error 22–3; statistical 21; symbolic 20; *see also* representation
modular construction 176–7
Mortenson, M. A. 142, 160
motion detectors 30
movement problem, in virtual reality 134–5
multi-criteria optimization 164, 166
multi-objective optimization 164, 166
multi-tasking 186

Native American languages 48
natural disasters 140
natural selection 165, 169
NBBJ 160, 178
needs, hierarchy of human 25
Neeley, Dennis 22
negative numbers 45
negotiated meaning 97
nested if conditions 42
Newell, Allen 78, 81, 84, 97
non-manifold geometry 62, *63*
non-photo-real (NPR) rendering 67
non-uniform rational B-spline (NURBS) surfaces 60–1, *60*, 112

Norman, Don 86
NPR *see* non-photo-real (NPR) rendering
numbers 45–7, 107; binary 37, *37*, 45; floating point 39, 45–7; negative 45
NURBS (non-uniform rational B-spline) surfaces 60–1, *60*, 112

object graphics *see* vector graphics
objective functions 14, 80, 164, 166
objects 40
object snaps 55, 57, 130
object viewing models 65
occupancy sensors 30, 178
office automation software 67
open-source software 119, 151
open standards 32
operating manuals 71
operating systems 36
optimization 14; parametric 70; Pareto 164, 166
over-the-shoulder tools 132
OVG Real Estate 175

paint graphics *see* raster graphics
paint programs 53–4, 57, 67–8, 113, 118
pair work 99, 142
Palladio's villas 26, 95
Palo Alto Research Center (PARC) 128, 135
parallel projection 66
parameters 41, 96–7
parametric design 70, 71, 119, 189
parametric graphic elements 68
parametric optimization 70
Pareto, Vilfredo 166
Pareto fronts 164, 166
Pareto optimization 164, 166
pattern languages 3, 94
performance goals 80
performative architecture 142
personalization 176
perspective projection 66
PHIDIAS 93
Phong shading 66
photon-mapping 66
photorealistic images 66–7, 146
physical models 132–3

pixels 53, 54, 113
plain text format 36, 49, 50
planar graphs 62
Plato 78, 91, 172
poché 113
point clouds 58, 62
Poirier, Erik A. 31
population growth 4
PostScript 57
power management systems 179
power outlets 30, 179
practice, changing 139–51; construction process changes 32, 142–3; and continuous learning 151; and data 143–6, *145*, 147–8; decline of typical conditions 148–9; energy simulation 149–50; ethics 140; knowledge loss 144, *145*; MacLeamy Curve 139, *140*, 160; pair work 99, 142; productivity challenge 141–2; research and development 150–1
Prairie-School Houses (Wright) 95
pre-fabrication 176
primitives 54, 55, 57, 108; and closure 111–12; and complexity 112; high- and low-level 108–10, *109*, 111
printers 49
problem solving *see* design problems; solutions
problem-space search 90
procedural model of design 6
productivity gains, missing 27–9, *28*, 141–2, 176
projection 66, 132
prototyping 170
pseudo-code 40
Purcell, A. T. 88
puzzle making 80–1, 87, 97, 98

Queen Anne houses 95

radiosity 66, 67
random access memory (RAM) 8, 38
raster graphics 53–4, *56*, 57, 113, 118
raster output devices 49
ray casting 66
ray-tracing 64

read-only memory (ROM) 36
reality simulation 67
real numbers 39, 45–7
real-time systems 71
refinement 14
rendering 52, 58, 105, 185, 190; b-rep models 59, 60; graphics pipeline 57, 64–7, *65*; graphics processing units (GPUs) 66; non-photo-real (NPR) 67; photorealistic images 66–7, 146; wireframe 59
repetitive execution 43
representation 8–9, *9*, 104–25; affordances and task 110; and analytical power 107–8; of buildings 113–15, *114*; capturing design rationale 118–19; challenges to single-model concept 120–1; closure and complexity 111–12; and cognition 106–10; conversions between 63–4, *64*, 121–4; designing in code 119; drawing coordination problem 112; and embedded knowledge 108–10, *109*; emergence 117–18; high- and low-level primitives 108–10, *109*, 111; mass customization 14, 119–20, 129; parametric design 119; physical models 132–3; round-tripping problem 124–5; shape grammars 95, 118, 119, 168; of spatial connectivity 115–17, *116*; topology 115–17, *116*; *see also* 2D graphics; 3D graphics; drawing; models; standard representations
resolution 53
RhinoScript 119
Rittel, Horst 78, 80, 97, 164
ROM (read-only memory) 36
Roman numerals 107
round off errors 22–3
round-tripping problem 124–5, 157–9, *158*
Ruskin, John 6

scale 89, 99
scale models 21
scale problem, in virtual reality 134
scanning 132
Scheer, David 139, 190
schematic drawings 105

Schön, Donald 6, 84, 88, 184
science of design 6, 164, 185
scientific notation 45
scripting 9, 39, 70, 110, 119, 189; *see also* computer programming
scripts 44
Seattle 179
Seattle Art Museum 148
secondary storage 35–6
selection: design as 91, 144; natural 165, 169
self-conscious design, 90
self-directed learning 188
sensors 30, 176, 177–8, 180
sequentiality 38
shading algorithms 60, 66
shadows 66
shape grammars 95, 118, 119, 168
short-term memory 85, 106
SIGGRAPH metafile projects 57
Simon, Herbert 6, 77, 78, 79, 81, 97, 106, 164, 185
simulation 149–50, 170
single-point-of-contact sensors 177, 178
situated action 78, 97–8, 164
sketchbooks 92
sketches 105
sketching software 53–4, 57, 67–8, 113, 118
Sketchpad program 11, 39, 127
Skidmore Owens and Merrill (SOM) 150, 171
smart buildings 5, 29–30, 177–9
smart construction kits 132
smartphone apps 178, 179
Smith, Lillian 149–50
smoothing algorithms 60, 61, 66
snaps 55, 57, 130
soap-bubble diagrams 62, *63*, 115, 116
social action, design as 97–8
software/tools 6, 7–9, *9*; analysis tools 70; application programming interfaces (APIs) 189; automated analysis tools 70; building operations 71, 147, 177–9; clash detection 32, 69, 143; and cognition 98–100; cognitive and social side-effects of 99–100, 189–91; drafting

11, 54–7, 58, 113, 129–31; in education 185; fabrication systems 14, 70–1, 91, 119–20, 127, 129, 142, 143, 148–9; file-format obsolescence 148; interoperability 12, 31–2, 144; office automation 67; open-source 119, 151; parametric design 70, 71, 119, 189; real-time systems 71; sketching 53–4, 57, 67–8, 113, 118; *see also* 2D graphics; 3D graphics; algorithms; building information modeling (BIM) software; CAD systems
solar power 179
solid models 59, 61–2, 63–4, *64*
solutions 81–2, 162–73; by analogy 90, 164, 170; and creativity 171–2; directed search 168; direct measurement 169–70; evaluation of 169–71; existence of 164; genetic algorithms 70, 165, 168–9; hill climbing strategy 165, 167–8; incremental improvement 164–5; by inspection 166–7; objective functions 14, 80, 164, 166; Pareto optimization 164, 166; prototyping 170; selecting design moves 168–9; simulation 170; state spaces 1, 12, 79–80, 163, 164–5, 167–9; stopping rules 14, 81–2, 171; subdivision approach 91–2, 162; trade-offs 163
space networks 116
space syntax 5, 94, 117
spatial connectivity 115–17, *116*
Squiggle 146
standardized digital product descriptions 145
standard representations 44–9; ASCII 37, 47–8; end-of-line 48–9; fonts 49; glyphs 49; numbers 45–7; scientific notation 45; text 37, 47–8; Unicode 48; units of measure 46
Stanford Research Institute (SRI) 127
Stanford University 28, 168
state spaces 1, 12, 79–80, 163, 164–5, 167–9
statistical models 21
steady-state assumptions 20

Stiny, George 95
stopping rules 14, 81–2, 171
Street Bump project, Boston 179
strings, text 39
Suchman, Lucy 97–8, 164
sunshades 178
surface reflectivity and texture 59
surface-shaded images 59
surveying techniques 142
Sutherland, Ivan 11, 39, 127
Sydney Opera House 148
symbolic models 20
symbols 55; in drawings 113–15, *114*

Talbott, Kyle 96, 100
task-oriented approach 34, 185, 186, 188, 189
Technology and the Future of Cities report 175, 176
tessellation 60, 66
test data 22
text 37, 39, 47–8
texture mapping 66
time-and-motion approach 91
tool neutrality 98
topology 115–17, *116*
Total Station electronic theodolites 142
trade-offs 163
travel, personal 92–3
Traveling Salesman Problem 13
Turing, Alan 10, 172
Turing machine 10
Turing Test 10, 172

ubiquitous computing 135–6
Unicode 48
units of measure 46
University College London 94
University of Hawai'i 48
Unwin, Simon 88
user experience design (UxD) 110
utility functions *see* objective functions
Utzon, Jørn 148
UxD (user experience design) 110

variables 40
vector graphics 53, 54–7, *56*, 58, 113, 115

view frustum 65–6, *65*
viewing models 65
virtuality 10, 35–6
virtual movement 134–5
virtual reality (VR) 133, 134–5, 180–1
virtual travel 92
viruses, computer 38
visual programming extensions 44
Vitruvius 5
voice interfaces 133
voxels (volumetric elements) 58, *59*, 62
VR *see* virtual reality (VR)

water use monitoring 178
weather data 22
Webber, Melvin M. 78, 80, 97, 164

Western Wood Products Association 19–20
wicked problems 78, 97, 98, 164
Wilson, Duff 148
winged edge data structure 108
wireframe models 58–9, *59*, 63–4, *64*
working memory 85, 106
Wright, Frank Lloyd 89, 95

Xerox 128, 135

Yan, M. 96

zoning codes 31
zoom variation 99